The *H*-function with Applications in Statistics and Other Disciplines

A.M. MATHAI
McGill University

R.K. SAXENA
Jodhpur University and
McGill University

The *H*-function with Applications in Statistics and Other Disciplines

A HALSTED PRESS BOOK

JOHN WILEY & SONS
New York London Sydney Toronto

Published in the Western Hemisphere by
Halsted Press, a Division of
John Wiley & Sons, Inc., New York

Library of Congress Cataloging in Publication Data

Mathai, A M
The *H*-function with applications in statistics
and other disciplines.

"Halsted Press Book."
Bibliography: p.
Includes indexes.
1. H-function. 2. Mathematical statistics.
3. Matrices. I. Saxena, Ram Kishore, 1931–
II. Title.
QA353.H9M37 515′.55 78–8224
ISBN 0–470–26380–6

Printed in India by Prabhat Press, Meerut

To

Lalita Mathai

Shashi Saxena

Preface

This book deals with H-functions, known in the literature as generalized Mellin-Barnes functions or generalized G-functions or Fox's H-functions. All the recent developments on H-functions are given in this book with key results in the text and other results in the exercises at the end of each chapter. Applications of the results in statistics and other disciplines as well as functions of matrix argument are discussed. This book can be used as a text or a reference book.

The special features of this book are:

(1) Almost all of the materials are based on recent research papers and they are not available in any of the books in the field.

(2) Each chapter contains a list of exercises. These problems are taken from recent research papers.

(3) This is the first book which deals with applications of H-function in statistics and other disciplines. These applications arise naturally and are not theoretical applications.

(4) This is the first book which deals with special functions of the matrix argument. This area is a fast-growing area with vast potential of applicability especially in statistical problems.

(5) This is the first book which gives computable representations of H-functions. These are series expansions when the poles of the integrand are not restricted to be simple.

(6) A list of the various special functions of scalar argument and many other useful results are given in the appendix.

Chapter 1 introduces the function, gives the various elementary properties, recurrence formulae and expansions of the H-function. Some asymptotic expansions are also discussed here.

Chapter 2 deals with generalizations and integrals involving H-functions,

Integral transforms, integrals of products of H-functions and Appell's functions are also discussed.

Chapter 3 is mainly devoted to series representations for the H-function. Series representations in terms of Laguerre polynomials, Gegenbauer polynomials, Legendre's and associated functions, hypergeometric functions are all discussed here. Also a computable series representation for the general H-function is given in this chapter.

Chapter 4 is devoted to applications in statistics. After pointing out the general structures such as products, ratios and linear combinations of random variables where the H-functions come in naturally while dealing with statistical distributions, the text goes into specific applications such as in deriving the distributions of multivariate test statistics, introducing generalized probability laws, examining the structural set-up of probability laws and characterization problems.

Chapter 5 deals with functions of matrix arguments. The various elementary special functions, where the arguments are symmetric positive definite matrices, are discussed. Then higher functions such as the generalized hypergeometric functions, G-functions and the H-functions with matrix arguments are considered. Then some applications into statistics especially in the derivations of non-null distributions of multivariate test statistics are discussed. Three approaches are discussed, namely, through zonal polynomials, through generalized Laplace transform and through a generalized Mellin transform.

A detailed appendix, dealing with gamma and related functions, hypergeometric functions, special cases of the H-function, H-function expressed in terms of elementary functions and vice versa and hypergeometric functions of several variables, is given at the end of the book before presenting a bibliography covering the work on the topic until 1975.

Since the H-function is the most generalized special function a full discussion of its properties and special cases will need several volumes to give an integrated work covering all aspects. Hence what we have tried in this book is to look at the various topics where active research work is going on. These topics form the bases of the various chapters. Then from each topic some key results are taken and discussed in the text. These are results with medium difficulty and complication. Some elementary results and the more complicated results are given at the end of each chapter in the form of exercises with the aim of giving a more or less full coverage of the recent research outputs in these areas.

One aim of the book is to motivate the research workers and students in the area of special functions to search for applications of the vast resources of theoretical results available in the literature. We concentrated on developing the theory of special functions which could be directly applicable to other areas. We have found applications in the various branches of statistics. These are outlined in the chapter on applications.

The last chapter gives a summary of the developments on special functions of matrix argument. Here also we consider only the theoretical aspects which have direct applications. It is hoped that by introducing these topics, where active research work is still going on, it will close the gap between theory and applications as well as invigorate interdisciplinary research. An attempt is made to incorporate an extensive and up-to-date bibliography. But after completing the initial draft about one and a half years ago we were separated by individual commitments elsewhere and the bibliography could not be updated after that.

We would like to thank Miss Hildegard Schroeder for the hard job of typing the initial draft and the National Research Council of Canada for some financial assistance. The preparation of the final draft was done while the first author was on his sabbatic leave from McGill University. The first author would like to thank the following institutions for providing him with all types of facilities: Instituto de Matematica, Universidade Estadual de Campinas, São Paulo, Brazil, Centre for Mathematical Sciences, Trivandrum, Kerala, India and the Indian Statistical Institute, New Delhi. We would also like to thank Professors R.G. Buschman, R.S. Kushwaha, B.R. Bhonsle, C.B. Rathie, H.M. Srivastava, K.C. Gupta and others for sending reprints and preprints and their keen interest in the project.

A.M. Mathai
R.K. Saxena

Montreal
November 1977

Contents

CHAPTER 1

The *H*-function

The *H*-function is applicable in a number of problems arising in physical sciences, engineering and statistics. The importance of this function lies in the fact that nearly all the special functions occurring in applied mathematics and statistics are its special cases. Besides, the functions considered by Boersma (1962), Mittag-Leffler, generalized Bessél function due to Wright (1935), the generalization of the hypergeometric functions studied by Fox (1928) and Wright (1935a, 1940) are all special cases of the *H*-function. Except the function of Boersma (1962), the aforesaid functions cannot be obtained as special cases of the *G*-function of Meijer (1946); hence a study of the *H*-function will cover wider range than the *G*-function and gives deeper, more general and more useful results directly applicable in various problems of physical and biological sciences.

On account of the presence of the coefficients of s in the definition (1.1.1) of the *H*-function given in the next section, the results of the *H*-function are obtainable in a more compact form and without much difficulty. This is not the case with a *G*-function. Chapter 4 which deals with the applications of the *H*-function will reveal this fact.

The results of this chapter are based on the work of Fox (1961), Braaksma (1964), Gupta (1965), Gupta and Jain (1966, 1968a and 1969), Anandani (1969a and 1969b), Lawrynowicz (1969), Skibinski (1970), Oliver and Kalla (1971), Nair (1972 and 1973), Buschman (1974) and others.

1.1 DEFINITION OF THE *H*-FUNCTION

Mellin-Barnes type integrals have been studied by Pincherle in 1888, Barnes (1908) and Mellin (1910). Dixon and Ferrar (1936) have given the asymptotic expansion of general Mellin-Barnes type integrals. Also, see Erdélyi et al. (1953, p. 49) in this connection.

Functions close to an *H*-function occur in the study of the solutions of certain functional equations considered by Bochner (1958) and

Chandrasekharan and Narasimhan (1962).

In an attempt to unify and extend the existing results on symmetrical Fourier kernels, Fox (1961) has defined the H-function in terms of a general Mellin-Barnes type integral. He has also investigated the most general Fourier kernel associated with the H-function and obtained the asymptotic expansions of the kernel for large values of the argument, by following his earlier method (Fox, 1928).

It is not out of place to mention that symmetrical Fourier kernels are useful in characterization of probability density functions and in obtaining the solutions of certain dual integral equations. In this connection, the reader is referred to the work of Fox (1965, 1965a), Saxena (1967, 1967a), Mathai and Saxena (1969a) and Saxena and Kushwaha (1972, 1972a).

An H-function is defined in terms of a Mellin-Bernes type integral as follows:

$$H^{m,n}_{p,q}(z) = H^{m,n}_{p,q}\left[z \,\middle|\, \begin{matrix}(a_p, A_p)\\(b_q, B_q)\end{matrix}\right]$$

$$= H^{m,n}_{p,q}\left[z \,\middle|\, \begin{matrix}(a_1, A_1), (a_2, A_2), \ldots, (a_p, A_p)\\(b_1, B_1), (b_2, B_2), \ldots, (b_q, B_q)\end{matrix}\right]$$

$$= \frac{1}{2\pi i}\int_L \chi(s)\, z^s\, ds, \tag{1.1.1}$$

where $i = (-1)^{1/2}$, $z \neq 0$

and $$z^s = \exp\,[s \operatorname{Log}|z| + i \arg z], \tag{1.1.2}$$

in which $\operatorname{Log}|z|$ represents the natural logarithm of $|z|$ and $\arg z$ is not necessarily the principal value. An empty product is interpreted as unity. Here

$$\chi(s) = \frac{\prod_1^m \Gamma(b_j - B_j s) \prod_1^n \Gamma(1 - a_j + A_j s)}{\prod_{m+1}^q \Gamma(1 - b_j + B_j s) \prod_{n+1}^p \Gamma(a_j - A_j s)}, \tag{1.1.3}$$

where m, n, p and q are nonnegative integers such that $0 \leqslant n \leqslant p$, $1 \leqslant m \leqslant q$; $A_j\,(j = 1, \ldots, p)$, $B_j\,(j = 1, \ldots, q)$ are positive numbers; $a_j\,(j = 1, \ldots, p)$, $b_j\,(j = 1, \ldots, q)$ are complex numbers such that

$$A_j\,(b_h + \nu) \neq B_h\,(a_j - \lambda - 1) \tag{1.1.4}$$

for $\nu, \lambda = 0, 1, 2, \ldots;\; h = 1, \ldots, m;\; j = 1, \ldots, n.$

L is a contour separating the points

$$s = \left(\frac{b_j + \nu}{B_j}\right), \;(j = 1, \ldots, m;\; \nu = 0, 1, \ldots) \tag{1.1.5}$$

which are the poles of $\Gamma(b_j - B_j s)$ $(j = 1, \ldots, m)$, from the points

$$s = \left(\frac{a_j - \nu - 1}{A_j}\right), (j = 1, \ldots, n; \nu = 0, 1, \ldots), \tag{1.1.6}$$

which are the poles of $\Gamma(1 - a_j + A_j s)$, $(j = 1, \ldots, n)$.

The contour L exists on account of (1.1.4). These assumptions will be retained throughout.

In the contracted form the H-function in (1.1.1) will be denoted by one of the following notations:

$$H(z),\ H_{p,q}^{m,n}(z),\ H_{p,q}^{m,n}\left[z \middle| \begin{matrix}(a_p, A_p)\\(b_q, B_q)\end{matrix}\right].$$

The H-function is an analytic function of z and makes sense if the following existence conditions are satisfied.

CASE 1. For all $z \neq 0$ with $\mu > 0$. (1.1.7)

CASE 2. For $0 < |z| < \beta^{-1}$ with $\mu = 0$. (1.1.8)

Here
$$\mu = \sum_{j=1}^{q} B_j - \sum_{j=1}^{p} A_j \tag{1 1.9}$$

and
$$\beta = \prod_{j=1}^{p} A_j^{A_j} \prod_{j=1}^{q} B_j^{-B_j} \tag{1.1.10}$$

It does not depend on the choice of L. Due to the occurrence of the factor z^s in the integrand of (1.1.1) it is, in general, multiple-valued but one-valued on the Riemann surface of $\log z$.

REMARK. An extension of the definition of the H-function has been discussed by Skibinski (1970).

1.2 SOME IDENTITIES OF THE H-FUNCTION

This section deals with certain elementary properties of the H-function. Many authors have given various properties of this function and the work of Braaksma (1964), Gupta (1965), Gupta and Jain (1966, 1968a and 1969), Bajpai (1969a), Lawrynowicz (1969), Anandani (1969a, 1969b) and Skibinski (1970) will be discussed here.

The results of this section follow readily from the definition of the H-function (1.1.1) and hence no proofs are given here.

PROPERTY 1.2.1 The H-function is symmetric in the pairs $(a_1, A_1), \ldots,$

(a_n, A_n), likewise $(a_{n+1}, A_{n+1}), \ldots, (a_p, A_p)$; in $(b_1, B_1), \ldots, (b_m, B_m)$ and in $(b_{m+1}, B_{m+1}), \ldots, (b_q, B_q)$.

PROPERTY 1.2.2 If one of the (a_j, A_j) $(j=1, \ldots, n)$ is equal to one of the (b_j, B_j) $(j=m+1, \ldots, q)$ [or one of the (b_j, B_j) $(j=1, \ldots, m)$ is equal to one of the (a_j, A_j) $(j=n+1, \ldots, p)$], then the H-function reduces to one of the lower order, and p, q and n (or m) decrease by unity.

Thus we have the following reduction formula:

$$H^{m,n}_{p,q}\left[x \,\middle|\, \begin{matrix}(a_1, A_1), (a_2, A_2), \ldots, (a_p, A_p)\\ (b_1, B_1), (b_2, B_2), \ldots, (b_{q-1}, B_{q-1}), (a_1, A_1)\end{matrix}\right] \tag{1.2.1}$$

$$= H^{m,n-1}_{p-1,q-1}\left[x \,\middle|\, \begin{matrix}(a_2, A_2), \ldots, (a_p, A_p)\\ (b_1, B_1), \ldots, (b_{q-1}, B_{q-1})\end{matrix}\right]$$

provided $n \geqslant 1$ and $q > m$.

PROPERTY 1.2.3

$$H^{m,n}_{p,q}\left[x \,\middle|\, \begin{matrix}(a_p, A_p)\\ (b_q, B_q)\end{matrix}\right] = H^{n,m}_{q,p}\left[\frac{1}{x} \,\middle|\, \begin{matrix}(1-b_q, B_q)\\ (1-a_p, A_p)\end{matrix}\right] \tag{1.2.2}$$

This is an important property of the H-function because it enables us to transform an H-function with $\mu = \sum_{j=1}^{q} B_j - \sum_{j=1}^{p} A_j > 0$ and arg x to one with $\mu < 0$ and arg $(1/x)$ and vice versa.

PROPERTY 1.2.4

$$\frac{1}{k} H^{m,n}_{p,q}\left[x \,\middle|\, \begin{matrix}(a_p, A_p)\\ (b_q, B_q)\end{matrix}\right] \tag{1.2.3}$$

$$= H^{m,n}_{p,q}\left[x^k \,\middle|\, \begin{matrix}(a_p, kA_p)\\ (b_q, kB_q)\end{matrix}\right]$$

where $k > 0$.

PROPERTY 1.2.5

$$x^{\sigma} H^{m,n}_{p,q}\left[x \,\middle|\, \begin{matrix}(a_p, A_p)\\ (b_q, B_q)\end{matrix}\right] = H^{m,n}_{p,q}\left[x \,\middle|\, \begin{matrix}(a_p+\sigma A_p, A_p)\\ (b_q+\sigma B_q, B_q)\end{matrix}\right] \tag{1.2.4}$$

PROPERTY 1.2.6

$$H^{m,n+1}_{p+1,q+1}\left[z \,\middle|\, \begin{matrix}(0, \gamma), (a_p, A_p)\\ (b_q, B_q), (r, \gamma)\end{matrix}\right] \tag{1.2.5}$$

$$= (-1)^r H^{m+1,n}_{p+1,q+1}\left[z \,\middle|\, \begin{matrix}(a_p, A_p), (0, \gamma)\\ (r, \gamma), (b_q, B_q)\end{matrix}\right]$$

where $p \leqslant q$.

PROPERTY 1 2.7

$$H^{m+1,n}_{p+1,q+1}\left[z\left|\begin{matrix}(a_p, A_p), (1-r, \gamma)\\(1, \gamma), (b_q, B_q)\end{matrix}\right.\right] \tag{1.2.6}$$

$$= (-1)^r H^{m+1,n}_{p+1,q+1}\left[z\left|\begin{matrix}(1-r, \gamma), (a_p, A_p)\\(b_q, B_q), (1,\gamma)\end{matrix}\right.\right],$$

$p \leqslant q$.

NOTE. In the above results (1.2.2) to (1.2.6) the branches of the H-functions are suitably chosen.

PROPERTY 1.2.8

$$H^{m,n}_{p,q}\left[z\left|\begin{matrix}(a_p, A_p)\\(b_q, B_q)\end{matrix}\right.\right] = (2\pi)^{(1-t)\alpha}\, t^{\beta}$$

$$\times H^{tm,tn}_{tp,tq}\left[(zt^{-\mu})^t\left|\begin{matrix}(\Delta(t, a_p), A_p)\\(\Delta(t, b_q), B_q)\end{matrix}\right.\right], \tag{1.2.7}$$

where t is a positive integer,

$$\alpha = m + n - \frac{p}{2} - \frac{q}{2},\ \beta = \sum_1^q b_j - \sum_1^p a_j + \frac{p}{2} - \frac{q}{2} + 1$$

$\mu = \sum_1^q B_j - \sum_1^p A_j$ and $(\Delta(t, \delta_r), \gamma_r)$ stands for

$$\left(\frac{\delta_r}{t}, \gamma_r\right), \left(\frac{\delta_r + 1}{t}, \gamma_r\right), \ldots, \left(\frac{\delta_r + t - 1}{t}, \gamma_r\right).$$

The result (1.2.7) is given by Gupta and Jain (1966) and is called the multiplication formula for H-functions.

For similar results, see Gupta and Jain (1969).

1.3 DERIVATIVES OF THE *H*-FUNCTION

Lawrynowicz (1969) has given the following four formulae for the successive derivatives of the H-function.

$$\frac{d^r}{dz^r}\left\{z^{-(\gamma b_1/B_1)} H^{m,n}_{p,q}\left[z^{\gamma}\left|\begin{matrix}(a_1, A_1), \ldots, (a_p, A_p)\\(b_1, B_1), \ldots, (b_q, B_q)\end{matrix}\right.\right]\right\} \tag{1.3.1}$$

$$= \left(\frac{-\gamma}{B_1}\right)^r z^{-r-\gamma b_1/B_1}$$

$$\times H^{m,n}_{p,q}\left[z^{\gamma}\left|\begin{matrix}(a_1, A_1), \ldots, (a_p, A_p)\\(r + b_1, B_1), (b_2, B_2), \ldots, (b_q, B_q)\end{matrix}\right.\right]$$

where $r = 1, 2, \ldots,$; $\gamma = B_1$ for $n > 1$.

$$\frac{d^r}{dz^r}\left\{z^{-rb_q/B_q} H_{p,q}^{m,n}\left[z^\gamma \middle| \begin{matrix}(a_1, A_1), \ldots, (a_p, A_p)\\(b_1, B_1), \ldots, (b_q, B_q)\end{matrix}\right]\right\} \tag{1.3.2}$$

$$= \left(\frac{\gamma}{B_q}\right)^r z^{-r-\gamma b_q/B_q}$$

$$\times H_{p,q}^{m,n}\left[z^\gamma \middle| \begin{matrix}(a_1, A_1), \ldots, (a_p, A_p)\\(b_1, B_1), \ldots, (b_{q-1}, B_{q-1}), (r + b_q, B_q)\end{matrix}\right]$$

where $r = 1, 2, \ldots,$; $m < q$; $\gamma = B_q$ for $r > 1$.

$$\frac{d^r}{dz^r}\left\{z^{-\gamma(1-a_1)/A_1} H_{p,q}^{m,n}\left[z^{-\gamma} \middle| \begin{matrix}(a_1, A_1), \ldots, (a_p, A_p)\\(b_1, B_1), \ldots, (b_q, B_q)\end{matrix}\right]\right\} \tag{1.3.3}$$

$$= \left(-\frac{\gamma}{A_1}\right)^r z^{-r-\gamma(1-a_1)/A_1}$$

$$\times H_{p,q}^{m,n}\left[z^{-\gamma} \middle| \begin{matrix}(-r + a_1, A_1), (a_2, A_2), \ldots, (a_p, A_p)\\(b_1, B_1), \ldots, (b_q, B_q)\end{matrix}\right]$$

where $n > 0$; $r = 1, 2, \ldots,$; $\gamma = A_1$, for $r > 1$.

$$\frac{d^r}{dz^r}\left\{z^{-\gamma(1-a_p)/A_p} H_{p,q}^{m,n}\left[z^{-\gamma} \middle| \begin{matrix}(a_1, A_1), \ldots, (a_p, A_p)\\(b_1, B_1), \ldots, (b_q, B_q)\end{matrix}\right]\right\} \tag{1.3.4}$$

$$= \left(\frac{\gamma}{A_p}\right)^r z^{-r-\gamma(1-a_p)/A_p}$$

$$\times H_{p,q}^{m,n}\left[z^{-\gamma} \middle| \begin{matrix}(a_1, A_1), \ldots, (a_{p-1}, A_{p-1}), (-r + a_p, A_p)\\(b_1, B_1), \ldots, (b_q, B_q)\end{matrix}\right]$$

for $p > n$; $r = 1, 2, \ldots,$; $\gamma = A_p$ for $r > 1$.

The results (1.3.1) to (1.3.4) for $r = 1$ are immediate consequences of the differentiation formulae given by Anandani (1969a). These results also hold for $r = 2, 3, \ldots,$ by an appeal to the principle of mathematical induction.

REMARK. The results of Lawrynowicz cited above are in a compact form and are convenient for practical applications.

Nair (1972 and 1973) has given four formulae for the derivative of the *H*-function. His results are the extensions of the formulae proved by Gupta and Jain (1968a). One of the formulae proved by Nair (1972) is

$$\left(x\frac{d}{dx} - c_1\right) \ldots \left(x\frac{d}{dx} - c_r\right)$$

$$\times \left\{x^s H_{p,q}^{m,n}\left[zx^h \middle| \begin{matrix}(a_p, A_p)\\(b_q, B_q)\end{matrix}\right]\right\} \tag{1.3.5}$$

$$= x^s H^{m,n+r}_{p+r,q+r}\left[zx^h \,\middle|\, \begin{matrix}(c_r - s, h), (a_p, A_p)\\ (b_q, B_q), (c_r - s + 1, h)\end{matrix}\right]$$

where $h > 0$.

When $c_1 = c_2 = \ldots = c_r = 0$, (1.3.5) reduces to a formula due to Gupta and Jain (1968a, p. 191).

Oliver and Kalla (1971) have derived four differentiation formulae for the H-function which extend the results of Anandani (1970c), which itself are the generalizations of the results due to Goyal, A. N. and Goyal, G. K. (1967). One of the results proved by Oliver and Kalla is the following:

$$\frac{d^r}{dx^r} H^{m,n}_{p,q}\left[(cx+d)^h \,\middle|\, \begin{matrix}(a_p, A_p)\\ (b_q, B_q)\end{matrix}\right] \tag{1.3.6}$$

$$= \frac{c^r}{(cx+d)^r} H^{m,n+1}_{p+1,q+1}\left[(cx+d)^h \,\middle|\, \begin{matrix}(0, h), (a_p, A_p)\\ (b_q, B_q), (r, h)\end{matrix}\right]$$

where c and d are complex numbers and h is real and positive.

1.4 RECURRENCE FORMULAE FOR THE H-FUNCTION

Gupta (1965) has obtained four recurrence formulae for the H-function by the method of integral transform due to Meijer (1940 and 1941). One of his results is given below:

$$(a_1 - a_2) H^{m,n}_{p,q}\left[x \,\middle|\, \begin{matrix}(a_1, A_1), (a_2, A_1), (a_3, A_3), \ldots, (a_p, A_p)\\ (b_1, B_1), \ldots, (b_q, B_q)\end{matrix}\right] \tag{1.4.1}$$

$$= H^{m,n}_{p,q}\left[x \,\middle|\, \begin{matrix}(a_1, A_1), (a_2 - 1, A_1), (a_3, A_3), \ldots, (a_p, A_p)\\ (b_1, B_1), \ldots, (b_q, B_q)\end{matrix}\right]$$

$$- H^{m,n}_{p,q}\left[x \,\middle|\, \begin{matrix}(a_1 - 1, A_1), (a_2, A_1), (a_3, A_3), \ldots, (a_p, A_p)\\ (b_1, B_1), \ldots, (b_q, B_q)\end{matrix}\right]$$

where $n \geqslant 2$.

Anandani (1969) has given six recurrence relations for the H-function which follow as a consequence of the definition of the H-function (1.1.1). Two such results are enumerated below:

$$(b_1 A_1 - a_1 B_1 + B_1)\, H^{m,n}_{p,q}\left[x \,\middle|\, \begin{matrix}(a_p, A_p)\\ (b_q, B_q)\end{matrix}\right] \tag{1.4.2}$$

$$= B_1\; H^{m,n}_{p,q}\left[x \,\middle|\, \begin{matrix}(a_1 - 1, A_1), (a_2, A_2), \ldots, (a_p, A_p)\\ (b_1, B_1), \ldots, (b_q, B_q)\end{matrix}\right]$$

$$+ A_1\; H^{m,n}_{p,q}\left[x \,\middle|\, \begin{matrix}(a_1, A_1), \ldots, (a_p, A_p)\\ (1 + b_1, B_1), (b_2, B_2), \ldots, (b_q, B_q)\end{matrix}\right]$$

where $m, n \geqslant 1$.

$$(b_q A_1 - a_1 B_q + B_q)\, H^{m,n}_{p,q}\left[x \,\middle|\, \begin{matrix}(a_p, A_p)\\ (b_q, B_q)\end{matrix}\right] \tag{1.4.3}$$

$$= B_q \; H_{p,q}^{m,n}\left[x \,\middle|\, \begin{matrix} (a_1-1, A_1), (a_2, A_2), \ldots, (a_p, A_p) \\ (b_1, B_1), \ldots, (b_q, B_q) \end{matrix}\right]$$

$$- A_1 \; H_{p,q}^{m,n}\left[x \,\middle|\, \begin{matrix} (a_1, A_1), \ldots, (a_p, A_p) \\ (b_1, B_1), \ldots, (b_{q-1}, B_{q-1}), (b_q+1, B_q) \end{matrix}\right]$$

where $n \geqslant 1, 1 \leqslant m \leqslant q-1$.

For further results on recurrence relations of the H-function, see the work of Bora and Kalla (1971) and Jain (1967). A set of contiguous relations for the H-function are given by Buschman (1974a).

1.5 EXPANSION FORMULAE FOR THE H-FUNCTION

The four expansion formulae for the G-function due to Meijer (1941a) have been extended to H-functions by Lawrynowicz (1969) by using a method analogous to one adopted by Meijer (1941a). The results are the following:

Let m, n, p and q be nonnegative integers, such that $1 \leqslant m \leqslant q$, $0 \leqslant n \leqslant p$. Further let $A_j\,(j=1, \ldots, p)$ and $B_j\,(j=1, \ldots, q)$ be positive numbers and $a_j\,(j=1, \ldots, p)$ and $b_j\,(j=1, \ldots, q)$ be complex numbers satisfying the condition (1.1.4) and $\mu > 0$, where μ is defined in (1.1.9). Then if ω and η are complex numbers such that $\omega \neq 0$ and $\eta \neq 0$, then the following results hold:

$$H_{p,q}^{m,n}\left[\eta\omega \,\middle|\, \begin{matrix} (a_p, A_p) \\ (b_q, B_q) \end{matrix}\right] = \eta^{(b_1/B_1)} \sum_{r=0}^{\infty} \frac{(1-\eta^{1/B_1})^r}{r!} \tag{1.5.1}$$

$$\times H_{p,q}^{m,n}\left[\omega \,\middle|\, \begin{matrix} (a_1, A_1), \ldots, (a_p, A_p) \\ (r+b_1, B_1), (b_2, B_2), \ldots, (b_q, B_q) \end{matrix}\right]$$

where η is arbitrary for $m=1$ and for $m>1$, $|\eta^{1/B_1}-1|<1$; $\arg(\eta\omega) = B_1 \arg(\eta^{1/B_1}) + \arg\omega$ and $|\arg \eta^{1/B_1}| < \pi/2$.

$$H_{p,q}^{m,n}\left[\eta\omega \,\middle|\, \begin{matrix} (a_p, A_p) \\ (b_q, B_q) \end{matrix}\right] = \eta^{(b_q/B_q)} \sum_{r=0}^{\infty} \frac{(\eta^{1/B_q}-1)^r}{r!} \tag{1.5.2}$$

$$\times H_{p,q}^{m,n}\left[\omega \,\middle|\, \begin{matrix} (a_1, A_1), \ldots, (a_p, A_p) \\ (b_1, B_1), \ldots, (b_{q-1}, B_{q-1}), (r+b_q, B_q) \end{matrix}\right]$$

where $q > m$, $|\eta^{1/B_q}-1|<1$; $\arg(\eta\omega) = B_q \arg(\eta^{1/B_q}) + \arg\omega$ and $|\arg \eta^{1/B_q}| < \pi/2$.

$$H_{p,q}^{m,n}\left[\eta\omega \,\middle|\, \begin{matrix} (a_p, A_p) \\ (b_q, B_q) \end{matrix}\right] = \eta^{(a_1-1)/A_1} \tag{1.5.3}$$

$$\times \sum_{r=0}^{\infty} \frac{(1-\eta^{-1/A_1})^r}{r!} \; H_{p,q}^{m,n}\left[\omega \,\middle|\, \begin{matrix} (-r+a_1, A_1), (a_2, A_2), \ldots, (a_p, A_p) \\ (b_1, B_1), \ldots, (b_q, B_q) \end{matrix}\right]$$

where $n > 0$, $R(\eta^{1/A_1}) > \frac{1}{2}$; $\arg(\eta\omega) = A_1 \arg \eta^{1/A_1} + \arg \omega$ and $|\arg \eta^{1/A_1}| < \pi/2$.

$$H^{m,n}_{p,q}\left[\eta\omega \middle| \begin{matrix}(a_p, A_p)\\(b_q, B_q)\end{matrix}\right] = \eta^{(a_p-1)/A_p} \sum_{r=0}^{\infty} \frac{(\eta^{-1/A_p}-1)^r}{r!} \times H^{m,n}_{p,q}\left[\omega \middle| \begin{matrix}(a_1, A_1), \ldots, (a_{p-1}, A_{p-1}), (a_p - r, A_p)\\(b_1, B_1), \ldots, (b_q, B_q)\end{matrix}\right] \tag{1.5.4}$$

where $p > n$, $R(\eta^{1/A_p}) > \frac{1}{2}$; $\arg(\eta\omega) = A_p \arg \eta^{1/A_p} + \arg \omega$ and $|\arg \eta^{1/A_p}| < \pi/2$.

For further details the reader is referred to the original paper by Lawrynowicz (1969).

REMARK. For a theorem on analytic continuation of the extended definition of the H-function, see the work of Skibinski (1970). Analytic continuation of the H-function has also been discussed by Braaksma (1964, p. 280).

1.6 ASYMPTOTIC EXPANSIONS

The behaviour of the H-function for small and large values of the argument has been discussed by Braaksma (1964) in detail. In this section we enumerate some of his results which are useful in applied problems. In order to present the results, the following definitions will be used.

$$\alpha = \sum_{j=1}^{n} A_j - \sum_{j=n+1}^{p} A_j + \sum_{j=1}^{m} B_j - \sum_{j=m+1}^{q} B_j;$$

$$\beta = \prod_{j=1}^{p} (A_j)^{A_j} \prod_{j=1}^{q} (B_j)^{-B_j};$$

$$\gamma = \sum_{j=1}^{q} b_j - \sum_{j=1}^{p} a_j + \frac{p}{2} - \frac{q}{2};$$

$$\lambda = \sum_{j=1}^{m} B_j - \sum_{j=m+1}^{q} B_j - \sum_{j=1}^{p} A_j, \text{ and } \mu, = \sum_{j=1}^{q} B_j - \sum_{j=1}^{p} A_j.$$

According to Braaksma (1964, p. 278)

$$H^{m,n}_{p,q}(x) = 0(|x|^c) \text{ for small } x. \tag{1.6.1}$$

where $\mu \geqslant 0$ and $c = \min R(b_j/B_j)$ $(j = 1, \ldots, m)$; and

$$H^{m,n}_{p,q}(x) = 0(|x|^d) \text{ for large } x, \tag{1.6.2}$$

where $\mu \geqslant 0$, $\alpha > 0$, $|\arg x| < \alpha\pi/2$ and

$$d = \max R\left(\frac{a_j - 1}{A_j}\right) \quad (j = 1, \ldots, n).$$

For $n = 0$, the H-function vanishes exponentially for large x in certain cases. We have

$$H_{p,q}^{m,0}[x] \sim 0\{\exp(-\mu x^{1/\mu}\beta^{1/\mu})\, x^{1/\mu(\gamma+1/2)}\} \tag{1.6.3}$$

provided that $\lambda > 0$, $|\arg x| < \pi\lambda/2$ and $\mu > 0$.

1.7 SPECIAL CASES

The H-function, being in a generalized form, contains a vast number of analytic functions as special cases. These analytic functions appear in various problems arising in theoretical and applied branches of mathematics, statistics and engineering sciences.

In the first place, when

$$A_j = B_h = 1 \; (j = 1, \ldots, p;\; h = 1, \ldots, q)$$

it reduces to a Meijer's G-function. We have

$$H_{p,q}^{m,n}\left[x \,\middle|\, \begin{matrix}(a_p, 1)\\(b_q, 1)\end{matrix}\right] = G_{p,q}^{m,n}\left[x \,\middle|\, \begin{matrix}a_p\\b_q\end{matrix}\right] \tag{1.7.1}$$

A detailed account of Meijer's G-function and its applications can be found in the monograph by Mathai and Saxena (1973a).

The G-function itself is a generalization of a number of known special functions listed on pages 53–68 of the monograph by Mathai and Saxena (1973a). We are, therefore, not listing all those special cases of the H-function which follow from the G-function. However, we list here a few interesting special cases of the H-function which may be useful for the workers on integral transforms, special functions, applied statistics, engineering sciences and perturbation theory.

$$H_{0,1}^{1,0}\left[z \,\middle|\, \begin{matrix}-\\(b, B)\end{matrix}\right] = B^{-1} z^{b/B} \exp(-z^{1/B}) \tag{1.7.2}$$

$$H_{1,1}^{1,1}\left[z \,\middle|\, \begin{matrix}(1-\nu, 1)\\(0, 1)\end{matrix}\right] = \Gamma(\nu)(1+z)^{-\nu} = \Gamma(\nu)\,{}_1F_0(\nu; -z) \tag{1.7.3}$$

$$H_{0,2}^{1,0}\left[\frac{z^2}{4} \,\middle|\, \left(\frac{a+\nu}{2}, 1\right), \left(\frac{a-\nu}{2}, 1\right)\right] = \left(\frac{z}{2}\right)^a J_\nu(z), \tag{1.7.4}$$

where $J_\nu(z)$ is the ordinary Bessel function of the first kind.

$$H_{0,2}^{2,0}\left[\frac{z^2}{4}\middle|\left(\frac{a-\nu}{2},1\right),\left(\frac{a+\nu}{2},1\right)\right]=2^{1-a}z^a K_\nu(z), \qquad (1.7.5)$$

where $K_\nu(z)$ is the modified Bessel function of the second kind.

$$H_{1,2}^{2,0}\left[z\middle|\begin{matrix}(a-\lambda+1,1)\\(a+\mu+\frac{1}{2},1),(a-\mu+\frac{1}{2},1)\end{matrix}\right]=z^a e^{-z/2}W_{\lambda,\mu}(z), \quad (1.7.6)$$

where $W_{\lambda,\mu}(z)$ is a Whittaker function.

$$H_{2,2}^{1,2}\left[z\middle|\begin{matrix}(1-\lambda,1)\ (1-\mu,1)\\(0,1),(1-\nu,1)\end{matrix}\right]=\frac{\Gamma(\lambda)\,\Gamma(\mu)}{\Gamma(\nu)}\,{}_2F_1(\lambda,\mu;\nu;-z) \quad (1.7.7)$$

The following two special cases of an H-function cannot be obtained from a G-function.

$$H_{p,q+1}^{1,p}\left[z\middle|\begin{matrix}(1-a_p,A_p)\\(0,1),(1-b_q,B_q)\end{matrix}\right]=\sum_{r=0}^{\infty}\frac{\prod_{1}^{p}\Gamma(a_j+A_jr)\,\frac{(-z)^r}{r!}}{\prod_{1}^{q}\Gamma(b_j+B_jr)} \qquad (1.7.8)$$

$$={}_p\psi_q\left[\begin{matrix}(a_p,A_p)\\(b_q,B_q)\end{matrix};-z\right],$$

which is called Maitland's generalized hypergeometric function.

The above series has been studied in detail by Wright (1935a). An interesting particular case of (1.7.7) gives a relation between an H-function and a Maitland's generalized Bessel function $J_\nu^\mu(z)$. The result is:

$$H_{0,2}^{1,0}\left[z\middle|(0,1),(-\nu,\mu)\right]=\sum_{r=0}^{\infty}\frac{(-z)^r}{r!\,\Gamma(1+\nu+\mu r)}=J_\nu^\mu(z). \qquad (1.7.9)$$

This result (1.7.8) is given by Braaksma (1964, p. 279).

We have also

$$H_{q+1,p}^{p,1}\left[z\middle|\begin{matrix}(1,1),(b_q,1)\\(a_p,1)\end{matrix}\right]=E(p;a_r:q;b_s:z), \qquad (1.7.10)$$

where E denotes MacRobert's E-function (MacRobert, 1969).

The following special cases of the H-function occur in the study of certain statistical distributions.

$$H_{2,2}^{2,0}\left[z\middle|\begin{matrix}(\alpha_1+\beta_1-1,1),(\alpha_2+\beta_2-1,1)\\(\alpha_1-1,1),(\alpha_2-1,1)\end{matrix}\right]$$

$$=\frac{z^{\alpha_2-1}(1-z)^{\beta_1+\beta_2-1}}{\Gamma(\beta_1+\beta_2)}\,{}_2F_1\left(\begin{matrix}\alpha_2+\beta_2-\alpha_1.\ \beta_1;1-z\\\beta_1+\beta_2\end{matrix}\right)(|z|<1). \quad (1.7.11)$$

$$H_{1,1}^{1,0}\left[z\,\middle|\,\begin{matrix}(a+1/2,\,1)\\(a,\,1)\end{matrix}\right]=\pi^{-1/2}z^{a}(1-z)^{-1/2}\ (|z|<1) \qquad (1.7.12)$$

$$H_{2,2}^{2,0}\left(z\,\middle|\,\begin{matrix}(a+1/3,\,1),\,(a+2/3,\,1)\\(a,\,1),\,(a,\,1)\end{matrix}\right)$$
$$=(2\pi)^{1/2}z^{a-1}-2\,z^{a-1/2}\,\pi^{-1/2}\,{}_2F_1(1/2,\,1/2;\,3/2;\,z)$$
$$=\frac{1}{\Gamma(3/2)}\,z^{a-1}(1-z)^{1/2}\,{}_2F_1(1/2,\,1/2;\,3/2;\,1-z)\ (|z|<1). \qquad (1.7.13)$$

(logarithmic case)

EXERCISES

1.1 Prove that if $R(\delta)>0$, then

(i) $f(x;\,\delta,\alpha,\gamma,1)=2\,(\gamma x/\alpha)^{\delta/2}\,K_{\delta}[2\,(\alpha\gamma x)^{1/2}]$;

(ii) $f(x;\,\delta,\alpha,\gamma,-1)=\Gamma(\delta)\,(\alpha+\gamma x)^{-\delta},\ R(\alpha+\gamma x)>0$;

(iii) $f(x;\,\delta,\alpha,\gamma,-\tfrac{1}{2})=2^{1-\delta}\,\Gamma(2\delta)\,\alpha^{-\delta}\exp(\alpha^{-1}\gamma^{2}x/8)\,D_{-2\delta}[(2\alpha)^{-1/2}\gamma x]$;

(iv) $f(x;\,\delta,\alpha,\gamma,-2)=\Gamma(\delta)\,(2\gamma x)^{-\delta/2}\exp\left[-\dfrac{\alpha^{2}}{\delta\gamma x}\right]D_{-\delta}[\alpha\,(2\gamma x)^{-1/2}]$;

where $f(x;\,\delta,\alpha,\gamma,\phi)=\alpha^{-\delta}H_{0,2}^{2,0}[\alpha^{\phi}\gamma x\,|_{(\delta,\phi),\,(0,1)}]$.

(Buschman, 1974)

1.2 Prove that

$$B_1\,z^{-b_1}H_{p,q}^{1,n}\left[z^{B_1}\,\middle|\,\begin{matrix}(a_p,\,A_p)\\(b_q,\,B_q)\end{matrix}\right]$$
$$=\sum_{\nu=0}^{\infty}\frac{\prod_{1}^{n}\Gamma\left[(1-a_j+A_j\left(\frac{b_1+\nu}{B_1}\right)\right](-z)^{\nu}}{\nu!\prod_{2}^{q}\Gamma\left[1-b_j+B_j\left(\frac{b_1+\nu}{B_1}\right)\right]\prod_{n+1}^{p}\Gamma\left[a_j-A_j\frac{b_1+\nu}{B_1}\right]}$$

(Braaksma, 1964, p. 279)

1.3 Prove that

(i) $$z^{r}\frac{d^{r}}{dz^{r}}\left\{H_{p,q}^{m,n}\left[z^{\delta}\,\middle|\,\begin{matrix}(a_p,\,A_p)\\(b_q,\,B_q)\end{matrix}\right]\right\}$$
$$=H_{p+1,q+1}^{m,n+1}\left[z^{\delta}\,\middle|\,\begin{matrix}(0,\,\delta),\,(a_p,\,A_p)\\(b_q,\,B_q),\,(r,\,\delta)\end{matrix}\right]$$

(ii) $$z^r \frac{d^r}{dz^r}\left\{ H^{m,n}_{p,q}\left[z^{-\delta} \middle| \begin{matrix} (a_p, A_p) \\ (b_q, B_q) \end{matrix} \right]\right\}$$
$$= (-1)^r H^{m,n+1}_{p+1,q+1}\left[z^{-\delta} \middle| \begin{matrix} (1-r, \delta), (a_p, A_p) \\ (b_q, B_q), (1, \delta) \end{matrix} \right]$$

giving the conditions of validity of the result.

Hint. Use the formulae

$$z^r \frac{d^r}{dz^r}(z^{s\delta}) = \frac{\Gamma(1+s\delta)}{\Gamma(1+s\delta-r)} z^{s\delta}$$

and

$$z^r \frac{d^r}{dz^r}(z^{-s\delta}) = \frac{(-1)^r \Gamma(r+s\delta)}{\Gamma(s\delta)} z^{-s\delta}.$$

(Skibinski, 1970, p. 132)

Show that

$$\frac{d^r}{dz^r}\left\{ z^{\lambda} H^{m,n}_{p,q}\left(\beta z^{\delta} \middle| \begin{matrix} (a_p, A_p) \\ (b_q, B_q) \end{matrix} \right)\right\}$$
$$= z^{\lambda-r} H^{m,n+1}_{p+1,q+1}\left[\beta z^{\delta} \middle| \begin{matrix} (-\lambda, \delta), (a_p, A_p) \\ (b_q, B_q), (r-\lambda, \delta) \end{matrix} \right]$$

where (f_r, g_r) denotes the set of parameters $(f_1, g_1), \ldots, (f_r, g_r)$.

(Anandani, 1970)

1.4 Establish the following identities.

(i) $$H^{m,n+1}_{p+1,q+1}\left[z \middle| \begin{matrix} (\alpha, \delta), (a_p, A_p) \\ (b_q, B_q), (\alpha+r, \delta) \end{matrix} \right]$$
$$= (-1)^r H^{m+1,n}_{p+1,q+1}\left[z \middle| \begin{matrix} (a_p, A_p), (\alpha, \delta) \\ (\alpha+r, \delta), (b_q, B_q) \end{matrix} \right]$$

(Anandani, 1970, p. 191)

(ii) $$H^{4,0}_{2,4}\left[z \middle| \begin{matrix} (1/2+a, 1), (1/2-a, 1) \\ (0, 1), (1/2, 1), (b, 1), (-b, 1) \end{matrix} \right]$$
$$= \left(\frac{\pi}{2}\right)^{1/2} W_{a,b}(2z^{1/2})\, W_{-a,b}(2z^{1/2}),$$

where $W_{a,b}(z)$ and $W_{-a,b}(z)$ are Whittaker functions.

(iii) $$H^{m,n+2}_{p+2,q+2}\left[z \middle| \begin{matrix} (-\sigma, h), (\alpha-\sigma, h), (a_p, A_p) \\ (b_q, B_q), (\alpha-\sigma-\nu, h), (-1-\beta-\sigma-\nu, h) \end{matrix} \right]$$
$$= (-1)^{\nu} H^{m+1,n+1}_{p+2,q+2}\left[z \middle| \begin{matrix} (-\sigma, h), (a_p, A_p), (\alpha-\sigma, h) \\ (\alpha-\sigma+\nu, h), (b_q, B_q), (-1-\beta-\sigma-\nu, h) \end{matrix} \right]$$

(iv) $$H^{m+1,n}_{p+1,q+1}\left[x \middle| \begin{matrix} (a_p, A_p), (\alpha-\beta-1, h) \\ (\alpha-\beta, h), (b_q, B_q) \end{matrix} \right]$$

$$= H^{m+1,n}_{p+1,q+1}\left[x\,\middle|\,\begin{matrix}(a_p, A_p), (\alpha+1, h)\\ (\alpha+2, h), (b_q, B_q)\end{matrix}\right]$$

$$-(\beta+2)\, H^{m,n}_{p,q}\left[x\,\middle|\,\begin{matrix}(a_p, A_p)\\ (b_q, B_q)\end{matrix}\right].$$

(Anandani 1969, pp. 136–39)

1.5 Prove that

$$\left(\frac{d}{dx}.x - c_1\right)\cdots\left(\frac{d}{dx}.x - c_r\right)\left\{x^s H^{m,n}_{p,q}\left[zx^h\,\middle|\,\begin{matrix}(a_p, A_p)\\ (b_q, B_q)\end{matrix}\right]\right\}$$

$$= x^s H^{m,n+r}_{p+r,q+r}\left[zx^h\,\middle|\,\begin{matrix}(c_r - s - 1, h), (a_p, A_p)\\ (b_q, B_q), (c_r - s, h)\end{matrix}\right]$$

where $h > 0$ and the symbol $\frac{d}{dx}.x$ indicates that the function of x in front of it is first multiplied by x and then the product is differentiated with respect to x.

Hence deduce the following result

$$\left(\frac{d}{dx}.x - c\right)\left(\frac{d}{dx}.x - c + e\right)\cdots\left(\frac{d}{dx}.x - c + [r-1]e\right)$$

$$\times\left\{x^{se+c-1} H^{m,n}_{p,q}\left[zx^{he}\,\middle|\,\begin{matrix}(a_p, A_p)\\ (b_q, B_q)\end{matrix}\right]\right\}$$

$$= e^r x^{se+c-1} H^{m,n+1}_{p+1,q+1}\left[zx^{he}\,\middle|\,\begin{matrix}(1-r-s, h), (a_p, A_p)\\ (b_q, B_q), (1-s, h)\end{matrix}\right]$$

provided $e \neq 0$, $h > 0$.

(Nair, 1972)

1.6 Establish the following differentiation formulae.

(i) $$\frac{d^r}{dx^r} H^{m,n}_{p,q}\left[(cx+d)^h\,\middle|\,\begin{matrix}(a_p, A_p)\\ (b_q, B_q)\end{matrix}\right]$$

$$= \frac{(-c)^r}{(cx+d)^r} H^{m+1,n}_{p+1,q+1}\left[(cx+d)^h\,\middle|\,\begin{matrix}(a_p, A_p), (0, h)\\ (r, h), (b_q, B_q)\end{matrix}\right]$$

(ii) $$\frac{d^r}{dx^r} H^{m,n}_{p,q}\left[\frac{1}{(cx+d)^h}\,\middle|\,\begin{matrix}(a_p, A_p)\\ (b_q, B_q)\end{matrix}\right]$$

$$= \frac{c^r}{(cx+d)^r} H^{m,n+1}_{p+1,q+1}\left[\frac{1}{(cx+d)^h}\,\middle|\,\begin{matrix}(a_p, A_p), (1-r, h)\\ (1, h), (b_q, B_q)\end{matrix}\right]$$

(iii) $$\frac{d^r}{dx^r} H^{m,n}_{p,q}\left[\frac{1}{(cx+d)^h}\,\middle|\,\begin{matrix}(a_p, A_p)\\ (b_q, B_q)\end{matrix}\right]$$

$$= \frac{(-c)^r}{(cx+d)^r} H^{m,n+1}_{p+1,q+1}\left[\frac{1}{(cx+d)^h}\,\middle|\,\begin{matrix}(1-r, h), (a_p, A_p)\\ (b_q, B_q), (1, h)\end{matrix}\right]$$

where c and d are complex numbers, r is a positive integer and $h > 0$.

(Oliver and Kalla, 1971)

1.7 Prove the following results.

(i) $$H_{p,q}^{m,n}\left[z\lambda^{\sigma} \left| \begin{matrix} (a_1, \sigma), (a_2, A_2), \ldots, (a_{p-1}, A_{p-1}), (a_p, \mu\sigma) \\ (b_1, B_1), \ldots, (b_q, B_q) \end{matrix} \right. \right]$$

$$= \lambda^{a_1 - 1} \sum_{r=0}^{\infty} \left(\frac{1}{r!}\right)\left(1 - \frac{1}{\lambda}\right)^r$$

$$\times H_{p,q}^{m,n}\left[z \left| \begin{matrix} (a_1 - r, \sigma), (a_2, A_2), \ldots, (a_{p-1}, A_{p-1}) (a_p, \mu\sigma) \\ (b_1, B_1), \ldots, (b_q, B_q) \end{matrix} \right. \right]$$

where $1 \leqslant n \leqslant p - 1$, $\mu > 0$, $\sigma > 0$ and λ and z are complex numbers.

(ii) $$H_{p+1,q+1}^{m+1,n}$$

$$\left[x \left| \begin{matrix} (a_1, \sigma), (a_2, A_2), \ldots, (a_{p-1}, A_{p-1}), (a_p, \mu\sigma), (a_1 + \nu - n', \sigma) \\ (a_1 + \nu, \sigma), (b_1, B_1), \ldots, (b_q, B_q) \end{matrix} \right. \right]$$

$$= \sum_{r=0}^{n'} (-1)^r \, n'_{c_r} \left\{ \prod_{m'=0}^{n'-r-1} (\nu - m') \right\}$$

$$\times H_{p,q}^{m,n}\left[x \left| \begin{matrix} (a_1 - r, \sigma), (a_2, A_2), \ldots, (a_{p-1}, A_{p-1}), (a_p, \mu\sigma) \\ (b_1, B_1), \ldots, (b_q, B_q) \end{matrix} \right. \right]$$

where $1 \leqslant n \leqslant p - 1$, $\mu > 0$, n' is a nonnegative integer and

$$\prod_{m'=0}^{n'-r-1} (\nu - m') = 1, \qquad \text{when } r = n'$$

$$= \nu, \qquad \text{when } r = n' - 1$$

$$= \nu(\nu - 1) \text{ when } r = n' - 2$$

and so on.

(iii) $$(a_p - \mu a_1) H_{p,q}^{m,n}\left[x \left| \begin{matrix} (1 + a_1, \sigma), (a_2, A_2), \ldots, (a_{p-1}, A_{p-1}), (1 + a_p, \mu\sigma) \\ (b_1, B_1), \ldots, (b_q, B_q) \end{matrix} \right. \right]$$

$$= H_{p,q}^{m,n}\left[x \left| \begin{matrix} (1 + a_1, \sigma), (a_2, A_2), \ldots, (a_{p-1}, A_{p-1}), (a_p, \mu\sigma) \\ (b_1, B_1), \ldots, (b_q, B_q) \end{matrix} \right. \right]$$

$$+ \mu H_{p,q}^{m,n}\left[x \left| \begin{matrix} (a_1, \sigma), (a_2, A_2), \ldots, (a_{p-1}, A_{p-1}), (a_p + 1, \mu\sigma) \\ (b_1, B_1), \ldots, (b_q, B_q) \end{matrix} \right. \right]$$

where $1 \leqslant n \leqslant p - 1$ and $\mu > 0$.

(iv) $$H_{p+1,q+1}^{m+1,n}$$

$$\left[x \left| \begin{matrix} (1 + a_1, \sigma), (a_2, A_2), \ldots, (a_{p-1}, A_{p-1}), (a_p, \mu\sigma), (a_1 + \nu, \sigma) \\ (a_1 + \nu + 1, \sigma), (b_1, B_1), \ldots, (b_q, B_q) \end{matrix} \right. \right]$$

$$= \nu\, H_{p,q}^{m,n}\left[x \,\middle|\, \begin{matrix}(1+a_1, \sigma), (a_2, A_2), \ldots, (a_{p-1}, A_{p-1}), (a_p, \mu\sigma)\\ (b_1, B_1), \ldots, (b_q, B_q)\end{matrix}\right]$$

$$- H_{p,q}^{m,n}\left[x \,\middle|\, \begin{matrix}(a_1, \sigma), (a_2, A_2), \ldots, (a_{p-1}, A_{p-1}), (a_p, \mu\sigma)\\ (b_1, B_1), \ldots, (b_q, B_q)\end{matrix}\right]$$

where $1 \leqslant n \leqslant p-1$ and $\mu > 0$.

(v) $(a_p - \mu a_1)$

$$\times H_{p+1,q+1}^{m+1,n}\left[x \,\middle|\, \begin{matrix}(1+a_1, \sigma), (a_2, A_2), \ldots, (a_{p-1}, A_{p-1}), (1+a_p, \mu\sigma), (a_1+\nu, \sigma)\\ (1+a_1+\nu, \sigma), (b_1, B_1), \ldots, (b_q, B_q)\end{matrix}\right]$$

$$= \nu\, H_{p,q}^{m,n}\left[x \,\middle|\, \begin{matrix}(1+a_1, \sigma), (a_2, A_2), \ldots, (a_{p-1}, A_{p-1}), (a_p, \mu\sigma)\\ (b_1, B_1), \ldots, (b_q, B_q)\end{matrix}\right]$$

$$+ (\nu\mu - a_p + \mu a_1)\, H_{p,q}^{m,n}\left[x \,\middle|\, \begin{matrix}(a_1, \sigma), (a_2, A_2), \ldots, (a_{p-1}, A_{p-1}), (1+a_p, \mu\sigma)\\ (b_1, B_1), \ldots, (b_q, B_q)\end{matrix}\right]$$

where $1 \leqslant n \leqslant p-1$, $\mu > 0$.

(vi) $$x^{1-1/\mu} \frac{d}{dx} x^{(a_p-1)/\mu}\, H_{p,q}^{m,n}\left[zx^{-\sigma} \,\middle|\, \begin{matrix}(a_1, \sigma), (a_2, A_2), \ldots, (a_{p-1}, A_{p-1}), (a_p, \mu\sigma)\\ (b_1, B_1), \ldots, (b_q, B_q)\end{matrix}\right]$$

$$= \left(\frac{1}{\mu}\right)^r x^{(a_p-r-1)/\mu}\, H_{p,q}^{m,n}\left[zx^{-\sigma} \,\middle|\, \begin{matrix}(a_1, \sigma), (a_2, A_2), \ldots, (a_{p-1}, A_{p-1}), (a_p-r, \mu\sigma)\\ (b_1, B_1), \ldots, (b_q, B_q)\end{matrix}\right]$$

where $1 \leqslant n \leqslant p-1$ and $\mu > 0$.

(Srivastava and Gupta, 1970)

Hint. The above results can be proved with the help of the theory of generalized Hankel transform given by Kumar (1954, p. 191) and recurrence relations and certain infinite series expansions established by Kumar (1954, pp. 192, 195, 199 and 1955, p. 42) and the integral of products of two H-functions due to Jain and Gupta (1966).

1.8 Prove that

$$J'_\lambda(xy) = (xy)^{-\lambda/2} J_\lambda(2(xy)^{1/2}).$$

Also show that

(i) $J_\lambda^\mu(x) = 0(1)$ as $|x| \to 0$

and

(ii) $J_\lambda^\mu(x) = 0[x^{-k(\lambda+1/2)} \exp\{(\mu x)^k (\cos(\pi k))/(\mu k)\}]$,

where $k = \dfrac{1}{1+\mu}$.

(Wright, 1935, p. 258)

(iii) $J_{\lambda}^{\mu}(st) = \sum_{r=0}^{\infty} \frac{(1-s)^r}{r!} t^{\nu} J_{\lambda+\mu r}^{\mu}(t),$

where J_{λ}^{μ} is Wright's generalized Bessel function (Kumar, 1957).

1.9 Let

$$d(b_1, a_p - k) = \det \begin{bmatrix} b_1 & a_p - k \\ B_1 & A_p \end{bmatrix}$$

in which the first row of the determinant is written by our notation. The second row of the determinant is always to be completed with the appropriate A's and B's corresponding to the a's and b's of the first row. Further, we employ the notation $H[b_1 + 1]$ to denote the contiguous function in which b_1 is replaced by $b_1 + 1$, but with all other parameters left unchanged. Similar meanings hold for all other contiguous H-functions occurring in this problem. In the following results H will denote the H-function.

Prove the following relations of contiguity for the H-function.

$$A_p H[b_1 + 1] - B_1 [a_p - 1] = d(b_1, a_p - 1) H. \quad (1.9.1)$$

$$A_p H[a_1 - 1] + A_1 H[a_p - 1] = -d((a_1 - 1), a_p - 1) H. \quad (1.9.2)$$

$$B_q H[a_1 - 1] - A_1 H[b_q + 1] = -d(a_1 - 1, b_q) H. \quad (1.9.3)$$

$$B_q H[b_1 + 1] + B_1 H[b_q + 1] = d(b_1, b_q) H. \quad (1.9.4)$$

$$A_1 H[b_1 + 1] + B_1 H[a_1 - 1] = d(b_1, a_1 - 1) H. \quad (1.9.5)$$

$$B_q H[a_p - 1] + A_p H[b_q + 1] = d(a_p - 1, b_q) H. \quad (1.9.6)$$

$$B_2 H[b_1 + 1] - B_1 H[b_2 + 1] = d(b_1, b_2) H. \quad (1.9.7)$$

$$A_2 H[a_1 - 1] - A_1 H[a_2 - 1] = -d(a_1 - 1, a_2 - 1) H. \quad (1.9.8)$$

$$A_{p-1} H[a_p - 1] - A_p H[a_{p-1} - 1] = d(a_p - 1, a_{p-1} - 1) H. \quad (1.9.9)$$

$$B_{q-1} H[b_q + 1] - B_q H[b_{q-1} + 1] = -d(b_q, b_{q-1}) H. \quad (1.9.10)$$

$$d(a_p - 1, b_q) H[a_1 - 1] - d(b_q - a_1 - 1) H[a_p - 1] = -d(a_1 - 1, a_p - 1) H[b_q + 1]. \quad (1.9.11)$$

$$d(a_p - 1, b_q) H[b_1 + 1] + d(b_q, b_1) H[a_p - 1] = d(b_1, a_p - 1) H[b_q + 1]. \quad (1.9.12)$$

$$d(a_1-1,\ b_q)\,H[b_1+1]-d(b_q,\ b_1)\,H[a_1-1]$$
$$=d(b_1,\ a_1-1)\,H[b_q+1]. \tag{1.9.13}$$

$$d(a_1-1,\ a_p-1)\,H[b_1+1]-d(a_p-1,\ b_1)\,H[a_1-1]$$
$$=-\,d(b_1,a_1-1)\,H[a_p-1]. \tag{1.9.14}$$

$$d(b_2,\ b_3)\,H[b_1+1]+d(b_3,\ b_1)\,H(b_2+1)$$
$$=-\,d(b_1,\ b_2)\,H[b_3+1]. \tag{1.9.15}$$

$$d\,(a_2-1,\ a_3-1)\,H[a_1-1]+d(a_3-1,\ a_2-1)\,H[a_2-1]$$
$$=-\,d(a_1-1,\ a_2-1)\,H[a_3-1]. \tag{1.9.16}$$

$$d\,(a_{p-1}-1,\ a_{p-2}-1)\,H[a_p-1]+d\,(a_{p-2}-1,\ a_p-1)\,H[a_{p-1}-1]$$
$$=-\,d\,(a_p-1,\ a_{p-1}-1)\,H[a_{p-2}-1]. \tag{1.9.17}$$

$$d\,(b_{q-1},\ b_{q-2})\,H[b_q+1]+d\,(b_q-2,\ b_q)\,H[b_{q-1}+1]$$
$$=-\,d\,(b_q,b_{q-1})\,H[b_{q-2}+1]. \tag{1.9.18}$$

$$d\,(a_p-1,\ b_1)\,H[a_{p-1}-1]+d\,(b_1,\ a_{p-1}-1)\,H[a_p-1]$$
$$=-\,d(a_{p-1}-1,\ a_p-1)\,H[b_1+1]. \tag{1.9.19}$$

$$d\,(b_q,\ a_1-1)\,H[b_{q-1}+1]+d\,(a_1-1,\ b_{q-1})\,H[b_q+1]$$
$$=-\,d\,(b_{q-1},\ b_q)\,H[a_1-1]. \tag{1.9.20}$$

$$d\,(a_2-1,\ a_p-1)\,H[a_1-1]+d\,(a_p-1,\ a_1-1)\,H[a_2-1]$$
$$=d(a_1-1,\ a_2-1)\,H[a_p-1]. \tag{1.9.21}$$

$$d\,(b_2,\ b_q)\,H[b_1+1]+d(b_q,\ b_1)\,H[b_2+1]$$
$$=d\,(b_1,\ b_2)\,H[b_q+1]. \tag{1.9.22}$$

$$d\,(a_{p-1}-1,\ a_1-1)\,H[a_p-1]+d\,(a_1-1,\ a_p-1)\,H[a_{p-1}-1]$$
$$=d\,(a_{p-1},\ a_{p-1}-1)\,H[a_1-1]. \tag{1.9.23}$$

$$d\,(b_{q-1},\ b_1)\,H[b_q+1]+d\,(b_1,\ b_q)\,H[b_{q-1}+1]$$
$$=d\,(b_q,\ b_{q-1})\,H[b_1+1]. \tag{1.9.24}$$

$$d\,(a_2-1,\ b_q)\,H[a_1-1]+d\,(b_q,\ a_1-1)\,H[a_2-1]$$
$$=-\,d\,(a_1-1,\ a_2-1)\,H[b_q+1]. \tag{1.9.25}$$

$$d\,(b_2,\ a_{p-1})\,H[b_1+1]+d\,(a_{p-1},\ b_1)\,H[b_2+1]$$
$$=-\,d\,(b_1,\ b_2)\,H[a_p-1]. \tag{1.9.26}$$

$$d(a_2-1, b_1)\, H[a_1-1] + d(b_1, a_1-1)\, H[a_2-1] = d(a_1-1, a_2-1)\, H[b_1+1]. \qquad (1.9.27)$$

$$d(b_2, a_1-1)\, H[b_1+1] + d(a_1-1, b_1)\, H[b_2+1] = d(b_1, b_2)\, H[a_1-1]. \qquad (1.9.28)$$

$$d(a_{p-1}-1, b_q)\, H[a_p-1] + d(b_q, a_p-1)\, H[a_{p-1}-1] = d(a_p-1, a_{p-1}-1)\, H[b_q+1]. \qquad (1.9.29)$$

$$d(b_{q-1}, a_p-1)\, H[b_q+1] + d(a_p-1, b_q)\, H[b_{q-1}+1] = d(b_q, b_{q-1})\, H[a_p-1]. \qquad (1.9.30)$$

(Buschman, 1972)

Hint. First establish the basic relations (1.9.1) and (1.9.2) given above and then derive all the others from two of them and using the transformation formula (1.2.2).

1.10 Establish the following results associated with the Mellin transforms of the partial derivatives of the H-function with respect to their parameters.

(i) $M\left\{\frac{\partial}{\partial b_1} H_{p,q}^{m,n}(x)\right\} = \chi(-s)\, \psi(b_1 + B_1 s),\ m > 0;$

(ii) $M\left\{\frac{\partial}{\partial a_1} H_{p,q}^{m,n}(x)\right\} = -\chi(-s)\, \psi(1 - a_1 - A_1 s),\ n > 0;$

(iii) $M\left\{\frac{\partial}{\partial a_p} H_{p,q}(x)\right\} = -\chi(-s)\, \psi(a_p + A_p s),\ n < p;$

(iv) $M\left\{\frac{\partial}{\partial b_q} H_{p,q}^{m,n}(x)\right\} = \chi(-s)\, \psi(1 - b_q - B_q s),\ m < q;$

(v) $M\left\{\frac{\partial}{\partial B_1} H_{p,q}^{m,n}(x)\right\} = s\chi(-s)\, \psi(b_1 + B_1 s),\ m > 0;$

(vi) $M\left\{\frac{\partial}{\partial A_1} H_{p,q}^{m,n}(x)\right\} = -s\chi(-s)\, \psi(1 - a_1 - A_1 s),\ n > 0;$

(vii) $M\left\{\frac{\partial}{\partial A_p} H_{p,q}^{m,n}(x)\right\} = -s\chi(-s)\, \psi(a_p + A_p s),\ n < p;$

(viii) $M\left\{\frac{\partial}{\partial B_q} H_{p,q}^{m,n}(x)\right\} = s\chi(-s)\, \psi(1 - b_q - B_q s),\ m < q,$

where M denotes the Mellin transform, ψ is the psi-function and $\chi(s)$ is given in (1.1.3) (Buschman 1974, p. 151)

CHAPTER 2

Generalized *H*-function and Integrals of *H*-functions

This chapter deals with integrals involving *H*-functions. There are a number of papers on integrals of *H*-functions by many workers notably by Gupta K.C. (1965, 1966), Gupta K.C. and Jain U.C. (1966), Sharma O.P. (1965, 1966, 1968, 1972), Anandani (1968, 1969, 1969d, 1969e, 1970, 1970b, 1970j, 1971, 1971b, 1971d, 1973, 1973a and 1973d), Gupta and Olkha (1969), Shah (1969a, 1969b, 1969d, 1972d), Bajpai (1969, 1969a–1969e, 1969–70, 1970, 1970a, 1970b), Goyal, S.P. (1970, 1971a), Saxena R.K. (1971a), Banerji and Saxena (1971, 1973), Rathie, P.N. (1967) and others. Since most of the work on integrals of *H*-functions is similar to that of *G*-function of Meijer (1946), which can be found in the monograph by Mathai and Saxena (1973a), we will give mostly those results which are not discussed in our monograph.

In this chapter we have derived an integral involving products of three *H*-functions of different arguments evaluated by Saxena, R.K. (1971) in terms of *H*-function of two variables defined in Section 2.2. This integral is of very general character and yields, as special cases, a number of integral transforms such as Mellin transform, Laplace transform, Meijer transform, etc. of the product of two *H*-functions of different arguments. Integrals of *H*-functions, when the integration is performed with respect to a parameter, due to Nair (1973a) and Samar (1973) are also given. Finite integrals of the products of *H*-functions and exponential functions due to Bajpai (1970) and Saxena (1971a) are discussed. Finally, this chapter ends with the evaluation of an integral involving product of an associated Legendre function and the *H*-function, evaluated by Singh and Varma (1972) by the use of finite difference operator *E*, and integrals of generalized Laguerre polynomials and the *H*-function given by Shah (1969a).

2.1 DEFINITIONS OF APPELL'S FUNCTIONS AND KAMPÉ DE FERIÉT'S DOUBLE HYPERGEOMETRIC FUNCTION

In this section we give the definitions of Appell's functions F_1, F_2, F_3 and F_4 and Kampé de Feriét's function of two variables.

Kashyap (1966) has used the Appell's function F_2 in deriving the solution of certain problems of double-ended queues with bulk service and limiting waiting space. An account of priority queues is given in a monograph by Jaiswal (1968).

Bailey (1933, 1934), Burchnall (1942), Saxena, R.K. (1966c) and Srivastava, H.M. (1973) and other workers have discussed some cases of the reducibility of Appell's function F_4. Certain integral representations for Appell's function F_4, in terms of integrals of products of Bessel functions of different arguments, are given by Saxena (1964a, 1966).

A detailed account of the theory of Appell's functions can be found in a monograph by Appell and Kampé de Feriét (1926), Bailey (1935), Slater (1961) and Rainville (1965). Kampé de Feriét's function of two variables is discussed in detail by Appell and Kampé de Feriét (1926). Integrals associated with a Kampé de Feriét's function and Gauss's hypergeometric function have been evaluated by Srivastava and Saran (1967) and Saxena, R.K. (1970). Expansion formulae for Kampé de Feriét function of two variables are developed by Verma, A. (1966, 1966a), Ragab (1973) and others. A generalization of Kampé de Feriét's double hypergeometric function is introduced by Srivastava and Daoust (1969).

The definitions of the Appell's functions are as follows:

$$F_1(a, b, b', c; x, y) = \sum_{m=0}^{\infty} \sum_{n=0}^{\infty} \frac{(a)_{m+n} (b)_m (b')_n}{(c)_{m+n}\, m!\, n!} x^m y^n, \tag{2.1.1}$$

where, for convergence, $|x| < 1$ and $|y| < 1$.

$$F_2(a, b, b', c, c'; x, y) = \sum_{m=0}^{\infty} \sum_{n=0}^{\infty} \frac{(a)_{m+n} (b)_m (b')_n}{(c)_m (c')_n\, m!\, n!} x^m y^n \tag{2.1.2}$$

where, for convergence, $|x| + |y| < 1$.

$$F_3(a, a', b, b', c; x, y) = \sum_{m=0}^{\infty} \sum_{n=0}^{\infty} \frac{(a)_m (a')_n (b)_m (b')_n}{(c)_{m+n}\, m!\, n!} x^m y^n \tag{2.1.3}$$

where, for convergence, $|x| < 1$, $|y| < 1$.

$$F_4(a, b, c, c'; x, y) = \sum_{m=0}^{\infty} \sum_{n=0}^{\infty} \frac{(a)_{m+n} (b)_{m+n}}{(c)_m (c')_n\, m!\, n!} x^m y^n \tag{2.1.4}$$

where, for convergence, $|x|^{1/2}+|y|^{1/2}<1$.

$$F\left[\begin{array}{c|cc|c} \mu & & (\alpha_\mu) & \\ \nu & (\beta_\nu) & ; \; (\beta'_\nu) & x \\ \rho & & (\gamma_\rho) & \\ \sigma & (\delta_\sigma) & ; \; (\delta'_\sigma) & y \end{array}\right] = \sum_{m=0}^{\infty}\sum_{n=0}^{\infty} \frac{(\alpha_\mu)_{m+n}(\beta_\nu)_m(\beta'_\nu)_n}{m!\,n!\,(\gamma_\rho)_{m+n}(\delta_\sigma)_m(\delta'_\sigma)_n} x^m y^n, \tag{2.1.5}$$

where $(a_A)_n$ stands for the parameters

$$(a_1)_n, (a_2)_n, \ldots, (a_A)_n$$

and

$$(a)_n = \frac{\Gamma(a+n)}{\Gamma(a)} = a(a+1)\ldots(a+n-1);\ (a)_0 = 1.$$

The series (2.1.5) converges absolutely for all complex values of x and y provided that $\mu+\nu<\rho+\sigma+1$ or $\mu+\nu=\rho+\sigma+1$ and $|x|+|y|<\min(1, 2^{\rho-\mu+1})$, (see Ragab, 1973).

REMARK. Certain recurrence formulae for the Kampé de Feriét's function are obtained by Srivastava, A. and Gupta K. C. (1971). On relations connecting G-function of two variables and Kampé de Feriét function, see the work of Agarwal, R.P. (1965). A new analytic continuation of Appell's function F_2 is recently given by Sud, K. and Wright, L.E. (1976).

2.2 A GENERALIZED *H*-FUNCTION

G-function of two variables given by Sharma, B. L. (1965) and Agarwal, R. P. (1965) arose from an attempt to generalize the functions of two variables defined in the preceding section. Since the *H*-function defined by Fox (1961) does not appear to be a special case of this function, several workers have introduced further extensions of this function and called them *H*-functions of two variables. The work of Pathak (1970) Verma, R. U. (1971) and Munot and Kalla (1971) may be mentioned in this direction. The definitions of the generalized *H*-function given by the aforesaid authors are the same. The importance of this function derives largely due to the fact that it includes, as special cases, a single *H*-function or product of two *H*-functions and most of the known functions of one and two variables, e.g. Meijer's *G*-function, MacRobert's *E*-function, *G*-function of two variables, Appell's functions F_1, F_2, F_3 and F_4 and the Whittaker functions of two variables, etc.

The generalized *H*-function is defined in terms of a double

Mellin-Barnes type integral in the following manner.

$$H\begin{bmatrix} x \\ y \end{bmatrix} = H^{L,N,N_1,M,M_1}_{E,(A:C),F,(B:D)}\left[\begin{array}{c|c} & (e_E,\theta_E) \\ x & (a_A,\alpha_A);(c_C,\gamma_C) \\ & (f_F,\phi_F) \\ y & (b_B,\beta_B);(d_D,\delta_D) \end{array}\right]$$

$$= H^{L,N,N_1,M,M_1}_{E,(A:C),F,(B:D)}\left[\begin{array}{c|c} x & (e_1,\theta_1),\ldots,(e_E,\theta_E) \\ & (a_1,A_1),\ldots,(a_A,\alpha_A);(c_1,\gamma_1),\ldots,(c_C,\gamma_C) \\ y & (f_1,\phi_1),\ldots,(f_F,\phi_F) \\ & (b_1,\beta_1),\ldots,(b_B,\beta_B);(d_1,\delta_1),\ldots,(d_D,\delta_D) \end{array}\right] \tag{2.2.1}$$

$$= -\frac{1}{4\pi^2}\int_{-i\infty}^{+i\infty}\int_{-i\infty}^{+i\infty} \chi_1(\xi)\,\chi_2(\eta)\,\chi_3(\xi+\eta)\,x^{\xi}y^{\eta}\,d\xi\,d\eta$$

where an empty product is interpreted as unity,

$$\chi_1(\xi) = \frac{\prod_1^M \Gamma(b_j-\beta_j\xi)\ \prod_1^N \Gamma(1-a_j+\alpha_j\xi)}{\prod_{M+1}^B \Gamma(1-b_j+\beta_j\xi)\ \prod_{N+1}^A \Gamma(a_j-\alpha_j\xi)} \tag{2.2.2}$$

$$\chi_2(\eta) = \frac{\prod_1^{M_1} \Gamma(d_j-\eta\delta_j)\ \prod_1^{N_1} \Gamma(1-c_j+\eta\gamma_j)}{\prod_{M_1+1}^D \Gamma(1-d_j+\eta\delta_j)\ \prod_{N_1+1}^C \Gamma(c_j-\eta\gamma_j)} \tag{2.2.3}$$

and

$$\chi_3(\rho) = \frac{\prod_1^L \Gamma(e_j+\rho\theta_j)}{\prod_{L+1}^E \Gamma(1-e_j-\rho\theta_j)\ \prod_1^F \Gamma(f_j+\rho\phi_j)} \tag{2.2.4}$$

The following assumptions are made:

(i) $0 \leqslant N \leqslant A,\ 1 \leqslant M \leqslant B,\ 0 \leqslant N_1 \leqslant C,\ 1 \leqslant M_1 \leqslant D,\ 0 \leqslant L \leqslant E.$

(ii) $L, M, N, M_1, N_1, A, B, C, D$ and E are nonnegative integers.

(iii) The sequence of parameters (a_A), (b_B), (c_C), (d_D), (e_E) and (f_F) are complex numbers and (α_A), (β_B), (γ_C), (δ_D), (θ_E) and (ϕ_F) are all real and positive numbers.

(iv) The sequence of parameters (a_A), (b_B), (c_C), (d_D), (e_E), (f_F), (α_A), (β_B), (γ_C), (δ_D), (θ_E) and (ϕ_F) are such that the poles of the integrand are simple.

(v) The paths of integration are indented, if necessary, in such a manner, that all the poles of $\Gamma(b_j - \beta_j\xi)$, $(j = 1, \ldots, M)$, $\Gamma(d_j - \delta_j\eta)$ $(j = 1, \ldots, M_1)$ lie to the right and those of $\Gamma(1 - a_j + \alpha_j\xi)$, $(j = 1, \ldots, N)$, $\Gamma(1 - c_j + \gamma_j)$, $(j = 1, \ldots, N_1)$ and $\Gamma[e_j + (\xi + \eta)\theta_j]$, $(j = 1, \ldots, L)$ lie to the left of the imaginary axis. The parameters are assumed to be such that this is possible.

The integral (2.1.1) converges absolutely if

$$\rho_j > 0, \ (j = 1, 2); \tag{2.2.5}$$

$$|\arg x| < \frac{\pi\rho_1}{2} \quad \text{and} \quad |\arg \eta| < \frac{\pi\rho_2}{2}, \tag{2.2.6}$$

where

$$\rho_1 = \sum_1^M \beta_j - \sum_{M+1}^B \beta_j + \sum_1^N \alpha_j - \sum_{N+1}^A \alpha_j + \sum_1^L \theta_j - \sum_{L+1}^E \theta_j - \sum_1^F \phi_j > 0; \tag{2.2.7}$$

$$\rho_2 = \sum_1^{M_1} \delta_j - \sum_{M_1+1}^D \delta_j + \sum_1^{N_1} \gamma_j - \sum_{N_1+1}^C \gamma_j + \sum_1^L \theta_j - \sum_{L+1}^E \theta_j - \sum_1^F \phi_j > 0. \tag{2.2.8}$$

For $L = 0$, the function $H\begin{bmatrix} x \\ y \end{bmatrix}$ will be denoted by

$$H^{(1)}\begin{bmatrix} x \\ y \end{bmatrix}$$

This function has the following property which is not enjoyed by $H\begin{bmatrix} x \\ y \end{bmatrix}$ and follows obviously by making changes in the variables.

$$H^{O,N,N_1,M,M_1}_{E,(A:C),F,(B:D)}\left[\begin{array}{c|c} & (e_1, \theta_1), \ldots, (e_E, \theta_E) \\ 1/x & (a_A, \alpha_A) ; (c_C, \gamma_C) \\ 1/y & (f_F, \phi_F) \\ & (b_B, \beta_B) ; (d_D, \delta_D) \end{array}\right] \tag{2.2.9}$$

$$= H^{O,M,M_1,N,N_1}_{F,(B:D),E,(A:C)}\left[\begin{array}{c|c} & (1 - f_F, \phi_F) \\ x & (1 - b_B, \beta_B) ; (1 - d_D, \delta_D) \\ y & (1 - e_1, \theta_1), \ldots, (1 - e_E, \theta_E) \\ & (1 - a_A, \alpha_A) ; (1 - c_C, \gamma_C) \end{array}\right]$$

This is an analogue of the formula (1.2.2) for the H-function.

If $k > 0$, we then obtain the identity

$$k^2\, H^{L,N,N_1,M,M_1}_{E,(A:C),F,(B:D)}\left[\begin{array}{c|c} x^k & (e_E, k\theta_E) \\ & (a_A, k\alpha_A); (c_C, k\gamma_C) \\ y^k & (f_F, k\phi_F) \\ & (b_B, k\beta_B); (d_D, k\delta_D) \end{array}\right] \tag{2.2.10}$$

$$= H^{L,N,N_1,M,M_1}_{E,(A:C),F,(B:D)}\left[\begin{array}{c|c} x & (e_E, \theta_E) \\ & (a_A, \alpha_A); (c_C, \gamma_C) \\ & (f_F, \phi_F) \\ y & (b_B, \beta_B); (d_D, \delta_D) \end{array}\right]$$

which has been pointed out by Munot and Kalla (1971). The function defined by (2.2.1) will either be called a generalized *H*-function or *H*-function of two variables.

REMARK. Mittal, P.K. and Gupta, K.C. (1972) have discussed a slightly modified form of the definition (2.2.1) which immediately follows from it by an obvious change in the parameters and variables. Goyal, G. K. (1971) has, however, given a slightly different definition of the *H*-function of two variables.

2.3 SOME INTERESTING SPECIAL CASES

The following special cases are worth mentioning. These results have been obtained by comparing the definition of the generalized *H*-function with other special functions with which the relation is to be determined. This is not an exhaustive list. Further results can be found by using the aforesaid procedure.

(i) A relation between the generalized *H*-function and *G*-function of two variables is

$$H^{L,N,N_1,M,M_1}_{E,(A:C),F,(B:D)}\left[\begin{array}{c|c} x & (e_E, 1) \\ & (a_A, 1);\ (c_C, 1) \\ y & (f_F, 1) \\ & (b_B, 1),\ (d_D, 1) \end{array}\right] \tag{2.3.1}$$

$$= G^{L,N,N_1,M,M_1}_{E,(A:C),F,(B:D)}\left[\begin{array}{c|c} x & (e_E) \\ & (a_A);\ (c_C) \\ y & (f_F) \\ & (b_B);\ (d_D) \end{array}\right]$$

A detailed account of G-function of two variables can be found in the works of Sharma, B.L. (1965) and Agarwal, R.P. (1965, 1970).

Integrals and series expansions of this function are developed by Sharma (1966, 1967, 1967a, 1968, 1968a, 1968b, 1969, 1971, 1971a, 1972), Abiodun and Sharma (1971 and 1973), Shah (1970, 1970a, 1971, 1971a, 1971b, 1971c, 1971d, 1971e) and several others. Integrals involving G-function of two variables are given by Srivastava, H.M. and Joshi, C.M. (1968, 1969), Verma, R.U. (1966, 1966a), Goyal (1970), Vyas and Saxena (1973), Srivastava, H.M. and Singhal, J.P. (1968, 1969) and others. Series expansions for this function are also developed by Verma, A. (1966, 1966a), Verma, R.U. (1967, 1969/1970, 1970, 1970a), Goyal, A.N. and Sharma, S. (1971, 1971a) and others.

A computable representation for the G-function of two variables, when the upper parameters differ by an integer, has been given by Mathai and Saxena (1972a). If we further specialize the parameters in (2.3.1), we obtain

$$H^{\mu,\nu,\nu,1,1}_{\mu,(\nu;\nu),\rho,(\sigma+1;\sigma+1)}\left[\begin{array}{c|c} -x & \begin{array}{c}(\alpha_\mu,1)\\(1-\beta_\nu,1);\ (1-\beta'_\nu,1)\end{array}\\ -y & \begin{array}{c}(\gamma_\rho,1)\\(0,1),\{1-(\delta_\sigma),1\};\ (0,1),\{1-(\delta'_\sigma),1\}\end{array}\end{array}\right]$$

$$=\frac{\prod\limits_{j=1}^{\mu}\Gamma(\alpha_j)\prod\limits_{j=1}^{\nu}\{\Gamma(\beta_j)\Gamma(\beta'_j)\}}{\prod\limits_{j=1}^{\rho}\Gamma(\gamma_j)\prod\limits_{j=1}^{\sigma}\{\Gamma(\delta_j)\Gamma(\delta'_j)\}} \tag{2.3.2}$$

$$\times F\left[\begin{array}{c|c|c}\mu & (\alpha_\mu) & \\ \nu & (\beta_\nu);\ (\beta'_\nu) & x\\ \rho & (\gamma_\rho) & y\\ \sigma & (\delta_\sigma);\ (\delta'_\sigma) & \end{array}\right]$$

$$=-\frac{1}{4\pi^2}\int_{-i\infty}^{+i\infty}\int_{-i\infty}^{+i\infty}\Gamma(-\xi)\Gamma(-\eta)$$

$$\times\frac{\prod\limits_{j=1}^{\mu}\Gamma(\alpha_j+\xi+\eta)\prod\limits_{j=1}^{\nu}\{\Gamma(\beta_j+\xi)\Gamma(\beta'_j+\eta)\}(-x)^\xi(-y)^\eta\,d\xi d\eta}{\prod\limits_{j=1}^{\rho}\Gamma(\gamma_j+\xi+\eta)\prod\limits_{j=1}^{\sigma}\{\Gamma(\delta_j+\xi)\Gamma(\delta'_j+\eta)\}} \tag{2.3.3}$$

where

$$F\left[\begin{array}{c|c} \mu & (\alpha_\mu) \\ \nu & (\beta_\nu);\quad (\beta'_\nu) \\ \rho & (\gamma_\rho) \\ \sigma & (\delta_\sigma);\quad (\delta'_\sigma) \end{array}\right]$$

is the Kampé de Feriét's function.

The following results easily follow from (2.3.3).

$$H^{1,1,1,1,1}_{1,[1:1],1,[1:1]}\left[\begin{array}{c|c} -x & (\alpha,1) \\ & (1-\beta,1);\ (1-\beta',1) \\ -y & (\gamma,1) \\ & (0,1)\quad;\quad (0,1) \end{array}\right] \tag{2.3.4}$$

$$= -\frac{1}{4\pi^2}\int_{-i\infty}^{+i\infty}\int_{-i\infty}^{+i\infty} \frac{\Gamma(-\xi)\,\Gamma(-\eta)\,\Gamma(\alpha+\xi+\eta)\,\Gamma(\beta+\xi)\,\Gamma(\beta'+\eta)}{\Gamma(\gamma+\xi+\eta)}$$

$$\times(-x)^{\xi}(-y)^{\eta}\,d\xi d\eta.$$

$$= \frac{\Gamma(\alpha)\,\Gamma(\beta)\,\Gamma(\beta')}{\Gamma(\gamma)}\,F_1(\alpha,\beta,\beta',\gamma;\,x,y).$$

$$H^{1,1,1,1,1}_{1,[1:1],0,[2:2]}\left[\begin{array}{c|c} -x & (\alpha,1) \\ & (1-\beta,1);\ (1-\beta',1) \\ -y & \text{---} \\ & (0,1),(1-\gamma,1);\,(0,1),(1-\gamma',1) \end{array}\right] \tag{2.3.5}$$

$$= -\frac{1}{4\pi^2}\int_{-i\infty}^{+i\infty}\int_{-i\infty}^{+i\infty} \frac{\Gamma(-\xi)\,\Gamma(-\eta)\,\Gamma(\alpha+\xi+\eta)\,\Gamma(\beta+\xi)\,\Gamma(\beta'+\eta)}{\Gamma(\gamma+\xi)\,\Gamma(\gamma'+\eta)}$$

$$\times(-x)^{\xi}(-y)^{\eta}\,d\xi d\eta$$

$$= \frac{\Gamma(\alpha)\,\Gamma(\beta)\,\Gamma(\beta')}{\Gamma(\gamma)\,\Gamma(\gamma')}\,F_2(\alpha,\beta,\beta',\gamma,\gamma';\,x,y).$$

$$H^{0,2,2,1,1}_{0,[2:2],1,[1:1]}\left[\begin{array}{c|c} & \text{---} \\ -x & (1-\alpha,1),(1-\beta,1);\,(1-\alpha',1),(1-\beta',1) \\ & (\gamma,1) \\ -y & (0,1)\quad;\quad(0,1) \end{array}\right] \tag{2.3.6}$$

$$= -\frac{1}{4\pi^2}\int_{-i\infty}^{+i\infty}\int_{-i\infty}^{+i\infty}\frac{\Gamma(-\xi)\,\Gamma(-\eta)\,\Gamma(\alpha+\xi)\,\Gamma(\beta+\xi)\,\Gamma(\alpha'+\eta)}{\Gamma(\gamma+\xi+\eta)}$$

$$\times\,\Gamma(\beta'+\eta)\,(-x)^{\xi}\,(-y)^{\eta}\,d\xi d\eta.$$

$$= \frac{\Gamma(\alpha)\,\Gamma(\beta)\,\Gamma(\alpha')\,\Gamma(\beta')}{\Gamma(\gamma)}\,F_3(\alpha,\alpha',\beta,\beta',\gamma;x,y)$$

$$H^{2,0,0,1,1}_{2,[0:0],0,[2:2]}\left[\begin{array}{c|c} -x & \begin{array}{c}(\alpha,1),(\beta,1)\\ -\;\;;\;\;-\\ \overline{}\\ (0,1),(1-\gamma,1);\;(0,1),(1-\gamma',1)\end{array}\\ -y & \end{array}\right] \tag{2.3.7}$$

$$= -\frac{1}{4\pi^2}\int_{-i\infty}^{+i\infty}\int_{-i\infty}^{+i\infty}\frac{\Gamma(\alpha+\xi+\eta)\,\Gamma(\beta+\xi+\eta)\,\Gamma(-\xi)\,\Gamma(-\eta)}{\Gamma(\gamma+\xi)\,\Gamma(\gamma'+\eta)}$$

$$\times\,(-x)^{\xi}\,(-y)^{\eta}\,d\xi d\eta$$

$$= \frac{\Gamma(\alpha)\,\Gamma(\beta)}{\Gamma(\gamma)\,\Gamma(\gamma')}\,F_4(\alpha,\beta,\gamma,\gamma';x,y)$$

(ii) If we set $L=E=F=C=N_1=0$, $M_1=D=1$, we find that

$$\lim_{y\to 0}\;H^{0,N,0,M,1}_{0,[A:0],0,[B:1]}\left[\begin{array}{c|c} x & \begin{array}{c}\cdots\cdots\\ (a_A,\alpha_A);\;\text{---}\\ \cdots\cdots\\ (b_B,\beta_B);\;(0,1)\end{array}\\ y & \end{array}\right] \tag{2.3.8}$$

$$= H^{M,N}_{A,B}\left[x\,\middle|\,\begin{array}{c}(a_A,\alpha_A)\\ (b_B,\beta_B)\end{array}\right].$$

Similarly

$$\lim_{y\to 0}\;H^{L,0,0,M,1}_{E,(0:0),0,(B:1)}\left[\begin{array}{c|c} x & \begin{array}{c}(e_E,\theta_E)\\ \overline{};\overline{}\\ (b_B,\beta_B);(0,1)\end{array}\\ y & \end{array}\right] \tag{2.3.9}$$

$$= H^{M,L}_{E,B}\left[x\,\middle|\,\begin{array}{c}(1-e_E,\theta_E)\\ (b_B,\beta_B)\end{array}\right]$$

(iii) On the other hand if we set $L=E=F=0$, the generalized H-function breaks up into a product of two H-functions, namely,

$$H^{0,N,N_1,M,M_1}_{0,(A:C),0,(B:D)}\left[\begin{array}{c|c} x & (a_A,\alpha_A);\,(c_C;\gamma_C)\\ y & (b_B,\beta_B);\,(d_D,\delta_D)\end{array}\right] \tag{2.3.10}$$

$$= H_{A,B}^{M,N}\left[x \,\middle|\, \begin{matrix}(a_A, \alpha_A)\\(b_B, \beta_B)\end{matrix}\right] H_{C,D}^{M_1,N_1}\left[y \,\middle|\, \begin{matrix}(c_C, \gamma_C)\\(d_D, \delta_D)\end{matrix}\right]$$

REMARK. The H-function of two variables discussed in this section has recently been generalized to H-function of n variables by Saxena, R. K. (1974).

2.4 ASYMPTOTIC EXPANSIONS

$$H\begin{bmatrix}x\\y\end{bmatrix} = 0\,(|x|^{\varepsilon_j}\,|y|^{\tau_K})\ (j = 1, \ldots, M;\, K = 1, \ldots, M_1) \quad (2.4.1)$$

for small values of x and y, where

$$\varepsilon_j = \min R\left(\frac{b_j}{\beta_j}\right), \quad \tau_K = \min R\left(\frac{d_K}{\delta_K}\right).$$

For large values of x and y, the associated function $H^{(1)}\begin{bmatrix}x\\y\end{bmatrix}$ has the behaviour

$$H^{(1)}\begin{bmatrix}x\\y\end{bmatrix} = 0\,(|x|^{\rho}\,|y|^{\sigma}) \quad (2.4.2)$$

where $\rho = \max R\left(\dfrac{a_j - 1}{\alpha_j}\right)\ (j = 1, \ldots, A)$

and $\sigma = \max R\left(\dfrac{c_j - 1}{\gamma_j}\right)\ (j = 1, \ldots. C)$

provided that the conditions (2.2.5) and (2.2.6) are satisfied.

Throughout this chapter the poles of the integrand of the H-functions of one and two variables discussed here are assumed to be simple.

2.5 AN INTEGRAL REPRESENTATION FOR THE GENERALIZED H-FUNCTION

THEOREM

$$c\int_0^\infty H_{A,B}^{M,N}\left[ax \,\middle|\, \begin{matrix}(a_A, \alpha_A)\\(b_B, \beta_B)\end{matrix}\right] H_{C,D}^{M_1,N_1}\left[bx \,\middle|\, \begin{matrix}(c_C, \gamma_C)\\(d_D, \delta_D)\end{matrix}\right]$$

$$\times\ H_{F,E}^{L,0}\left[cx \,\middle|\, \begin{matrix}(f_F - 1, \phi_F)\\(-e_E, \theta_E)\end{matrix}\right] dx$$

$$= H_{E,(A:C),F,(B:D)}^{L,N,N_1,M,M_1}\left[\begin{matrix}a/c\\b/c\end{matrix}\,\middle|\, x\right], \quad (2.5.1)$$

where χ represents the parameters

$$\begin{bmatrix} (\theta_E = e_E, \theta_E) \\ (a_A, \alpha_A); (c_C, \gamma_C) \\ (\phi_F + f_F - 1, \phi_F) \\ (b_B, \beta_B); (d_D, \delta_D) \end{bmatrix}$$

$$R\left[\min\left(\frac{b_j}{\beta_j}\right) + \min\left(\frac{d_k}{\delta_k}\right) - \min\left(\frac{e_h}{\theta_h}\right) + 1\right] > 0,$$

$j = 1, \ldots, M$; $k = 1, \ldots, M_1$ *and* $h = 1, 2, \ldots, L$; $\lambda_1, \lambda_2, \lambda_3 > 0$, $|\arg a| < \pi\lambda_1/2$, $|\arg b| < \pi\lambda_2/2$ *and* $|\arg c| < \pi\lambda_3/2$, *where*

$$\lambda_1 = \sum_1^M \beta_j - \sum_{M+1}^B \beta_j + \sum_1^N \alpha_j - \sum_{N+1}^A \alpha_j;$$

$$\lambda_2 = \sum_1^{M_1} \delta_j - \sum_{M_1+1}^D \delta_j + \sum_1^{N_1} \gamma_j - \sum_{N_1+1}^C \gamma_j;$$

and

$$\lambda_3 = \sum_1^L \theta_j - \sum_{L+1}^E \theta_j - \sum_1^F \phi_j;$$

$$\sum_1^A \alpha_j - \sum_1^B \beta_j \leqslant 0; \sum_1^C \gamma_j - \sum_1^D \delta_j \leqslant 0, \text{ and}$$

$$\sum_1^F \phi_j - \sum_1^E \theta_j \leqslant 0.$$

PROOF. On substituting the value of $H_{A,B}^{m,n}[ax]$ in terms of Mellin-Barnes integral (1.1.1) in the integrand of (2.5.1) and interchanging the order of integration, which is obviously permissible under the conditions stated with the result, the integral transforms into

$$\frac{c}{2\pi i}\int_{-i\infty}^{+i\infty} \chi_1(\xi) a^{\xi} \int_0^{\infty} x^{\xi} H_{C,D}^{M_1,N_1}\left[bx \middle| \begin{matrix} (c_C, \gamma_C) \\ (d_D, \delta_D) \end{matrix}\right]$$

$$\times H_{F,E}^{L,0}\left[cx \middle| \begin{matrix} (f_F - 1, \phi_F) \\ (-e_E, \theta_E) \end{matrix}\right] dx\, d\xi.$$

On evaluating the x-integral by means of the formula given by Gupta and Jain (1966), it gives us

$$\frac{1}{2\pi i}\int_{-i\infty}^{+i\infty} \chi_1(\xi)\left(\frac{a}{c}\right)^{\xi}$$

$$H^{M_1,N_1+L}_{C+E,F+D}\left[\frac{b}{c}\middle|\begin{matrix}(1+e_E-(1+\xi)\,\theta_E,\theta_E),(c_C,\gamma_C)\\(2-f_F-(1+\xi)\,\phi_F,\phi_F),(d_D,\delta_D)\end{matrix}\right]d\xi$$

$$=-\frac{1}{4\pi^2}\int_{-i\infty}^{+i\infty}\int_{-i\infty}^{+i\infty}\chi_1(\xi)\,\chi_2(\eta)$$

$$\times\frac{\prod_{1}^{L}\Gamma\,[\theta_j-e_j+(\xi+\eta)\,\theta_j]\left(\frac{a}{c}\right)^{\xi}\left(\frac{b}{c}\right)^{\eta}d\xi\,d\eta}{\prod_{L+1}^{E}\Gamma[1-\theta_j+e_j+(\xi+\eta)\,\theta_j]\ \prod_{1}^{F}\Gamma[-1+f_j+\phi_j+(\xi+\eta)\,\phi_j]}$$

and the result follows from (2.2.1).

2.6 INTEGRAL TRANSFORMS OF PRODUCT OF TWO *H*-FUNCTIONS

The following special cases of (2.5.1) are interesting because they provide us the various integral transforms of the product of two *H*-functions of different arguments.

(i) If we set $L=E=2$, $F=1$, $\theta_1=\theta_2=1$, $\phi_1=1$, $f_1=2-k+\rho$, $e_1=-\rho-m-1/2$, $e_2=m-\rho-1/2$, then on using the identity (1.7.6), we find that the Whittaker transform of the product of two *H*-functions is given by

$$\int_0^\infty x^{\rho}\,e^{-ax/2}\,W_{k,m}(ax)\;H^{M,N}_{A,B}\left[bx\middle|\begin{matrix}(a_A,\alpha_A)\\(b_B,\beta_B)\end{matrix}\right]$$

$$\times H^{M_1,N_1}_{C,D}\left[cx\middle|\begin{matrix}c_C,\gamma_C)\\d_D,\delta_D)\end{matrix}\right]dx \tag{2.6.1}$$

$$=a^{-\rho-1}\,H^{2,N,N_1,M,M_1}_{2,[A:C],1,[B:D]}\left[\begin{matrix}\frac{b}{a}\\ \\ \frac{c}{a}\end{matrix}\middle|\begin{matrix}(\rho+m+3/2,1),\ (\rho-m+3/2,1)\\(a_A,\alpha_A);\ (c_C,\gamma_C)\\(2-k+\rho,1)\\(b_B,\beta_B);\ (d_D,\delta_D)\end{matrix}\right]$$

where $R[\rho\pm m+\min(b_j/\beta_j)+\min(d_h/\delta_h)+3/2]>0$, $j=1,\ldots,M$; $h=1,\ldots,M_1$; $\lambda_1,\lambda_2>0$, $R(a)>0$, $|\arg b|<\pi\lambda_1/2$, $|\arg c|>\pi\lambda_2/2$, λ_1 and λ_2 are defined in (2.5.1), $\sum_1^A\alpha_j-\sum_1^B\beta_j\leqslant 0$ and $\sum_1^C\gamma_j-\sum_1^D\delta_j\leqslant 0$.

For $k=0$, $m=\frac{1}{2}$, (2.6.1) gives the Laplace transform of the product of two *H*-functions, namely

$$\int_0^\infty x^\rho e^{-ax} H_{A,B}^{M,N}\left[bx\left|\begin{matrix}(a_A, \alpha_A)\\(b_B, \beta_B)\end{matrix}\right.\right] H_{C,D}^{M_1,N_1}\left[cx\left|\begin{matrix}(c_C, \gamma_C)\\(d_D, \delta_D)\end{matrix}\right.\right] dx$$

$$= a^{-\rho-1} H_{1,[A:C],0,[B:D]}^{1,N,N_1,M,M_1}\left[\begin{matrix}\frac{b}{a}\\ \\ \frac{c}{a}\end{matrix}\left|\begin{matrix}(1+\rho, 1)\\(a_A, \alpha_A); (c_C, \gamma_C)\\ -\\(b_B, \beta_B); (d_D, \delta_D)\end{matrix}\right.\right] \tag{2.6.2}$$

where $R(\rho + \min(b_j/\beta_j) + \min(d_h/\delta_h) + 1) > 0$, $(j = 1, \ldots, M$ and $h = 1, \ldots, M_1)$, $R(a) > 0$, $\lambda_1, \lambda_2 > 0$, $|\arg b| < \pi\lambda_1/2$ and $|\arg c| < \pi\lambda_2/2$; $\sum_1^A \alpha_j - \sum_1^B \beta_j \leqslant 0$ and $\sum_1^C \gamma_{j'} - \sum_1^D \delta_j \leqslant 0$.

(ii) Putting $L = 1, E = 2, F = 0, \theta_1 = \theta_2 = 1, e_1 = (-\rho-\nu)/2, e_2 = (\nu-\rho)/2$ and using the identity (1.7.4), it is observed that the Hankel transform of the product of two H-functions can be read from the relation,

$$\int_0^\infty x^{\rho-1} J_\nu(ax)\, H_{A,B}^{M,N}\left[bx^2\left|\begin{matrix}(a_A, \alpha_A)\\(b_B, \beta_B)\end{matrix}\right.\right] H_{C,D}^{M_1,N_1}\left[cx^2\left|\begin{matrix}(c_C, \gamma_C)\\(d_D, \delta_D)\end{matrix}\right.\right] dx$$

$$= 2^{\rho-1} a^{-\rho} H_{2,[A:C],0,[B:D]}^{1,N,N_1,M,M_1}\left[\begin{matrix}\frac{4b}{a^2}\\ \\ \frac{4c}{a^2}\end{matrix}\left|\begin{matrix}\left(\frac{\rho+\nu}{2}, 1\right); \left(\frac{\rho-\nu}{2}, 1\right)\\(a_A, \alpha_A); (c_C, \gamma_C)\\ -\\(b_B, \beta_B); (d_D, \delta_D)\end{matrix}\right.\right] \tag{2.6.3}$$

where $R\left[\frac{\nu}{2} + \min\left(\frac{b_j}{\beta_j}\right) + \min\left(\frac{d_h}{\delta_h}\right) + \frac{\rho}{2}\right] > 0, (j=1, \ldots, M; h=1, \ldots, M_1)$, $a > 0$, $R\left[\frac{\rho}{2} + \max\left(\frac{a_j - 1}{\alpha_j}\right) + \max\left(\frac{c_h - 1}{\gamma_h}\right)\right] < \frac{3}{4}$ $(j = 1, \ldots, N;$ $h = 1, \ldots, N_1)$; $|\arg b| < \frac{\pi\lambda_1}{2}, |\arg c| < \frac{\pi\lambda_2}{2}$;

$$\sum_1^A \alpha_j - \sum_1^B \beta_j \leqslant 0, \sum_1^C \gamma_j - \sum_1^D \delta_j \leqslant 0, \lambda_1 > 0 \text{ and } \lambda_2 > 0.$$

On using (1.7.4) in (2.6.3), we obtain Bailey's formula for the integral of the product of three Bessel functions (Bailey, 1936).

(iii) Similarly, if we employ the identity (1.7.5) we obtain the Meijer's K-transform of the product of two H-functions:

$$\int_0^\infty x^{\rho-1} K_\nu(ax)\, H^{M,N}_{B}\left[bx^2 \middle| \begin{matrix}(a_A, \alpha_A)\\(b_B, \beta_B)\end{matrix}\right] H^{M_1,N_1}_{C,D}\left[cx^2 \middle| \begin{matrix}(c_C, \gamma_C)\\(d_D, \delta_D)\end{matrix}\right] dx$$

$$= 2^{\rho-2} a^{-\rho} H^{2,N,N_1,M,M_1}_{2[A:C],0,[B:D]}\left[\begin{matrix}\frac{4b}{a^2}\\ \\ \frac{4c}{a^2}\end{matrix}\middle| \begin{matrix}\left(\frac{\rho\pm\nu}{2}, 1\right)\\ (a_A, \alpha_A); -(c_C, \gamma_C)\\ (b_B, \beta_B); (d_D, \delta_D)\end{matrix}\right] \tag{2.6.4}$$

where $$R\left[\frac{\rho\pm\nu}{2} + \min\left(\frac{b_j}{\beta_j}\right) + \min\left(\frac{d_h}{\delta_h}\right)\right] > 0,$$

$(j = 1, \ldots, M; h = 1, \ldots, M_1)$, $R(a) > 0$; $\lambda_1 > 0, \lambda_2 > 0, |\arg b| < \frac{\pi\lambda_1}{2}$;

$$|\arg c| < \frac{\pi\lambda_2}{2}; \sum_1^A \alpha_j - \sum_1^B \beta_j \leqslant 0 \text{ and } \sum_1^C \gamma_j - \sum_1^D \delta_j \leqslant 0.$$

REMARK 1. The Mellin transform of a single H-function follows from the definition (1.1.1) in view of the well known Mellin inversion theorem. We have

$$\int_0^\infty x^{s-1} H^{m,n}_{p,q}\left[ax \middle| \begin{matrix}(a_p, A_p)\\(b_q, B_q)\end{matrix}\right] dx \tag{2.6.5}$$

$$= a^{-s} \frac{\prod_1^m \Gamma(b_j + B_j s) \prod_{j=1}^n \Gamma(1 - a_j - A_j s)}{\prod_{m+1}^q \Gamma(1 - b_j - B_j s) \prod_{n+1}^p \Gamma(a_j + A_j s)}$$

where $-\min\limits_{1\leqslant j\leqslant m} R(b_j/B_j) < R(s) < \frac{1}{A_j} - \max\limits_{1\leqslant j\leqslant n} R(a_j/A_j), |\arg a| < \frac{1}{2}\pi\lambda$;

$$\lambda = \sum_1^n A_j - \sum_{n+1}^p A_j + \sum_1^m B_j - \sum_{m+1}^q B_j > 0, \text{ and} \tag{2.6.6}$$

$$\mu = \sum_1^p A_j - \sum_1^q B_j \leqslant 0. \tag{2.6.7}$$

The Mellin transform of the product of two H-functions has been given by Gupta and Jain (1966) in the following form.

$$\int_0^\infty x^{s-1} H^{m,n}_{p,q}\left[zx^\sigma \middle| \begin{matrix}(a_p, A_p)\\(b_q, B_q)\end{matrix}\right] H^{\alpha,\beta}_{\gamma,\delta}\left[x \middle| \begin{matrix}(d_\gamma, D_\gamma)\\(e_\delta, E_\delta)\end{matrix}\right] dx$$

$$= H^{m+\beta,n+\alpha}_{p+\delta,q+\gamma}\left[z \,\middle|\, \begin{matrix}(1-e_\delta-sE_\delta,\ \sigma E_\delta),\ (a_p, A_p)\\ (b_m, B_m),\ (1-d_\gamma-sD_\gamma,\ \sigma D_\gamma),\ (b_{m+1}, B_{m+1}), \ldots, (b_q, B_q)\end{matrix}\right] \tag{2.6.8}$$

where $\sigma > 0,\ \lambda \leqslant 0,\ \mu > 0,\ |\arg z| < \frac{1}{2}\pi\lambda^*,$

$$R\left[s + \sigma \min\left(\frac{b_h}{B_h}\right) + \min \frac{e_j}{E_j}\right] > 0, \ (h = 1, \ldots m; j = 1, \ldots, \alpha),$$

$$R\left[s + \sigma \max\left(\frac{a_h - 1}{A_h}\right) + \max\left(\frac{d_j - 1}{D_j}\right)\right] < 0, \ (h=1, \ldots, n; j=1, \ldots, \beta);$$

$$\sum_1^\gamma D_j - \sum_1^\delta E_j \leqslant 0;$$

$$\overset{*}{\mu} = \sum_1^\beta D_j - \sum_{\beta+1}^\gamma D_j + \sum_1^\alpha E_j - \sum_{\alpha+1}^\delta E_j > 0,$$

and λ and μ are defined in (2.6.6) and (2.6.7), respectively.

For the application of this result in the theory of statistical distributions, see the work of Mathai and Saxena (1969).

In what follows, λ and μ will have the values according to equations (2.6.6) and (2.6.7), respectively.

REMARK 2. Integrals involving the product of three H-functions are also evaluated by Nair and Samar (1971) and Nath (1972).

REMARK 3. For a generalization of the integral (2.5.1), see the work of Banerji and Saxena (1973) and for similar other results see Saxena (1971b). For integrals associated with the generalized H-function and Gauss's hypergeometric functions, see the work of Bora and Kalla (1970). Integrals of H-function of two variables are also evaluated by Sharma, C.K. (1973), Sharma, C.K. and Gupta, P.M. (1972), Shah (1972g, 1973, 1973a, 1973b), Munot (1972), Munot and Kalla (1971) and several others.

REMARK 4. One and two dimensional integral transformations associated with H-function of two variables are defined and studied by Bora, Kalla and Saxena (1970), Saxena, Bora and Kalla (1971), De Gomez Lopez and Kalla (1973) and others. A formal solution of dual integral equations associated with H-function of two variables is obtained by Saxena and Sethi (1973b) by the application of certain operators of fractional integration. Certain operators of fractional integration associated with H-function are defined by Kalla (1969) which have been generalized by Saxena and Kumbhat (1973, 1974). Saxena (1973a) has established Abelian theorems for a distributional transform associated with an H-function.

REMARK 5. Integral transforms of product of any finite number o H-functions of different arguments may be obtained from a recent paper due to Saxena (1974) in terms of H-function of n variables, defined there.

REMARK 6. A convolution integral equation associated with an H-function, has been solved by Srivastava and Buschman (1974). By the application of the Laplace transform operator L and its inverse L^{-} a solution of an integral equation involving H-function is derived by Verma, R.U. (1974). An integral equation whose kernel has a Mellin-Barnes type integral representation is solved by Saxena (1966a). For further results on the solution of integral equation associated with an H-function, see the works of Srivastava, H.M. (1972a), Saxena, V.P. (1970), Saxena and Kumbhat (1974a, 1974b) and Pathak and Prasad (1972).

2.7 INTEGRATION OF THE H-FUNCTION WITH RESPECT TO A PARAMETER

The following integrals have been established by Nair (1973) by the application of a theorem due to Lowndes (1964). For the details of the proof of these integrals, the reader is referred to the original paper of Nair (1973).

$$\int_0^\infty \frac{\Gamma\left(\frac{1-a\pm it}{2}\right)}{\Gamma(\pm\, it)} S_{a,it}(2x^{1/2})\, H_{p,q+2}^{m+2,n}\left[y\,\middle|\,\begin{matrix}(a_p, A_p)\\(b\pm it/2, k), (b_q, B_q)\end{matrix}\right] dt \quad (2.7.1)$$

$$= \pi\, 2^{a+1}\, x^b\, H_{p+1,q+1}^{m+1,n+1}\left[\frac{y}{x^k}\,\middle|\,\begin{matrix}\left(\frac{1+2b-a}{2}, k\right), (a_p, A_p)\\ \left(\frac{1+2b-a}{2}, k\right), (b_q, B_q)\end{matrix}\right]$$

provided that

$$R(a) \leqslant 1,\ R\left(k\frac{b_j}{B_j} - a - b\right) > -\frac{5}{4},\ R\left(k\frac{b_j}{B_j} + \frac{a}{2} - b\right) > -\frac{1}{2},$$

$$R\left(b + k\frac{(1-a_h)}{A_h} > 0\right) \text{ for } h = 1, \ldots, n;\ j = 1, \ldots, m;$$

$$|\arg y| < \frac{\pi\lambda}{2},\ \lambda > 0,\ \mu \leqslant 0,\ i = (-1)^{1/2},$$

and $\Gamma(a \pm b)$ indicates the product of two gamma functions $\Gamma(a+b)$ $\Gamma(a-b)$.

$$\int_0^\infty \frac{\Gamma\left(\frac{a \pm it}{2}\right)}{\Gamma(\pm it)} H_{p,q+2}^{m+2,n}\left[2^\rho y \middle| \begin{matrix}(a_p, A_p)\\(b \pm it, \rho), (b_q, B_q)\end{matrix}\right] dt \tag{2.7.2}$$

$$= \pi^{1/2}\, 2^{1-a+b}\, H_{p,q+2}^{m+2,n}\left[y \middle| \begin{matrix}(a_p, A_p)\\(a+b, \rho), (b+\frac{1}{2}, \rho), (b_q, B_q)\end{matrix}\right]$$

where $R(a) > 0$, $R\left[b - \frac{\rho}{A_j}(a_j - 1)\right] > 0$, for $j = 1, \ldots, n$;

$$\lambda > 0, |\arg y| < \frac{\pi\lambda}{2}, \lambda > 0 \text{ and } \mu \leqslant 0.$$

On specializing the parameters in (2.7.2) the following results involving Whittaker functions can be deduced.

$$e^{-x/2} W_{k,m}(x) \tag{2.7.3}$$

$$= \frac{(4/x)^{m+1/2}}{4\Gamma(\frac{1}{2})\,\Gamma(\frac{1}{2} - k \pm m)} \int_0^\infty \frac{\left(\frac{1}{4} + m \pm \frac{it}{2}\right)(-m - k \pm it)}{\Gamma(\pm it)} \times W_{k+m+\frac{1}{2}, it}(2x)\, dt$$

where $R(k+m) < 0$ and $R(m) > -1/4$.

$$e^{x/2} W_{k,m}(x) = \frac{(\ /x)^{m+1/2}}{4\sqrt{\pi}} \int_0^\infty \frac{\Gamma(\frac{1}{4} + m \pm it)}{\Gamma(\pm it)} W_{k-m-\frac{1}{2}, it}(2x)\, dt$$

where $R(m) > -\frac{1}{4}$.

Samar (1973) has shown that

$$\int_0^\infty \frac{\Gamma\left(\frac{1}{2} - a \pm \frac{it}{2}\right)}{\Gamma(\pm it)} W_{a, it/2}(x)\, H_{p,q+2}^{m+2,n}\left[y \middle| \begin{matrix}(a_p, A_p)\\(b \pm it/2, k), (b_q, B_q)\end{matrix}\right] dt$$

$$= 4\pi x^{b+1/2} e^{-x/2} H_{p,q+1}^{m+1,n}\left[\frac{y}{x^k} \middle| \begin{matrix}(a_p, A_p)\\(\frac{1}{2} - a + b, k), (b_q, B_q)\end{matrix}\right]$$

where $R\left(k\frac{b_j}{B_j} - 2a - b\right) > -\frac{5}{4}$,

$$R\left(k\frac{b_j}{B_j} - a - b\right) > -1 \ (j = 1, \ldots, m);$$

$$R[b + k(1 - a_j)/A_j] > 0, \ (j = 1, \ldots, n); \ R(a) \leqslant \tfrac{1}{2}, \ R(x) > 0,$$

$$\lambda > 0, |\arg y| < \frac{\pi\lambda}{2} \text{ and } \mu \leqslant 0.$$

REMARK. On the integration of H-functions with respect to their parameters, see the works of Nair and Nambudiripad (1973), Anandani (1970b), Taxak (1971), Golas (1968) and Pendse (1970). Integration of products of generalized Legendre functions and H-function with respect to a parameter is discussed by Anandani (1970b and 1971d).

2.8 INTEGRALS INVOLVING *H*-FUNCTION AND EXPONENTIAL FUNCTIONS

The following integrals are proved by Bajpai (1970) with the help of the integral (Erdélyi et al., 1953, p. 12)

$$\int_{-\pi/2}^{\pi/2} (\cos\theta)^{\alpha-1} e^{i\beta\theta}\, d\theta = \frac{\pi\,\Gamma(\alpha)}{2^{\alpha-1}\,\Gamma\left(\frac{\alpha+\beta+1}{2}\right)\Gamma\left(\frac{\alpha-\beta+1}{2}\right)}$$

where $R(\alpha) > 0$.

$$\int_{-\pi/2}^{\pi/2} (\cos\theta)^{k+l-2} e^{i(k-l)\theta}\, H_{p,q}^{m,n}\left[z(e^{i\theta}\cos\theta)^{-h}\,\middle|\,\begin{matrix}(a_p, A_p)\\(b_q, B_q)\end{matrix}\right] d\theta \tag{2.8.1}$$

$$= \frac{\pi}{2^{k+l-2}\,\Gamma(l)}\, H_{p+1,q+1}^{m+1,n}\left[2^h z\,\middle|\,\begin{matrix}(a_p, A_p), (k, h)\\(k+l-1, h), (b_q, B_q)\end{matrix}\right]$$

where $R(k+l-ha_j/A_j) > 1-h/A_j$; $(j=1, \ldots, n)$, $h>0$, $\mu \leqslant 0$; $\lambda > 0$ and $|\arg z| < \pi\lambda/2$.

$$\int_{-\pi/2}^{\pi/2} (\cos\theta)^{k+l-2} e^{i(k-l)\theta}\, H_{p,q}^{m,n}\left[ze^{ih\theta}(\sec\theta)^h\,\middle|\,\begin{matrix}(a_p, A_p)\\(b_q, B_q)\end{matrix}\right] d\theta \tag{2.8.2}$$

$$= \frac{\pi}{2^{k+l-2}\,\Gamma(k)}\, H_{p+1,q+1}^{m+1,n}\left[2^h z\,\middle|\,\begin{matrix}(a_p, A_p), (l, h)\\(k+l-1, h), (b_q, B_q)\end{matrix}\right]$$

where $R(k+l-ha_j/A_j) > 1-h/A_j\ (j=1, \ldots, n)$; $h>0$, $\mu \leqslant 0$, $\lambda > 0$ and $|\arg z| < \pi\lambda/2$.

$$\int_{-\pi/2}^{\pi/2} (\cos\theta)^{k+l-2} e^{i(k-l)\theta}\, H_{p,q}^{m,n}\left[ze^{2ih\theta}\,\middle|\,\begin{matrix}(a_p, A_p)\\(b_q, B_q)\end{matrix}\right] d\theta \tag{2.8.3}$$

$$= \frac{\pi\,\Gamma(k+l-1)}{2^{k+l-2}}\, H_{p+1,q+1}^{m,n}\left[z\,\middle|\,\begin{matrix}(a_p, A_p), (l, h)\\(b_q, B_q), (1-k, h)\end{matrix}\right]$$

where $R(k+l) > 1$, $h>0$, $\lambda > 0$, $\mu \leqslant 0$ and $|\arg z| < \pi\lambda/2$.

$$\int_{-\pi/2}^{\pi/2} (\cos\theta)^{k+l-2} e^{i(k-l)\theta}\, H_{p,q}^{m,n}\left[z(\sec\theta)^{2h}\,\middle|\,\begin{matrix}(a_p, A_p)\\(b_q, B_q)\end{matrix}\right] d\theta \tag{2 8.4}$$

$$= \frac{\pi}{2^{k+l-2}} H^{m+1,n}_{p+2,q+1}\left[2^{2h} z \middle| \begin{matrix}(a_p, A_p), (k, h), (l, h)\\(k+l-1, 2h), (b_q, B_q)\end{matrix}\right]$$

where $R(k+l-2h\,a_j/A_j) > 1-2h/A_j$ $(j=1,\ldots,n)$; $h>0$, $\lambda>0$, $\mu\leqslant 0$ and $|\arg z| < \pi\lambda/2$.

Saxena (1971a) has established the following results with the help of the integral [Nielson, 1906, p. 158]

$$\int_0^{\pi} (\sin t)^{\alpha} e^{-\beta t}\, dt = \frac{\Gamma(1+\alpha)}{2^{\alpha}\,\Gamma\left(1+\dfrac{\alpha \pm i\beta}{2}\right)} e^{-\pi\beta/2},\ R(\alpha+1)>0.$$

$$\int_0^{\pi} (\sin\theta)^{\gamma-1} e^{-\delta\theta} H^{m,n}_{p,q}\left[z(\sin\theta)^{2h}\middle| \begin{matrix}(a_p, A_p)\\(b_q, B_q)\end{matrix}\right] d\theta \tag{2.8.5}$$

$$= \frac{\pi}{2^{\gamma-1}} \exp\left(-\frac{\pi\delta}{2}\right) H^{m,n+1}_{p+1,q+2}\left[\frac{z}{4^h}\middle| \begin{matrix}(1-\gamma, 2h), (a_p, A_p)\\(b_q, B_q), \left(\dfrac{1\pm i\delta-\gamma}{2}, h\right)\end{matrix}\right]$$

where $R(\gamma+2h\,b_j/B_j)>0$ $(j=1,\ldots,m)$; $\lambda>0$, $h>0$, $\mu\leqslant 0$ and $|\arg z| < \pi\lambda/2$.

$$\int_0^{\pi} (\sin\theta)^{\gamma-1} e^{-\delta\theta} H^{m,n}_{p,q}\left[z e^{i2h\theta}\middle| \begin{matrix}(a_p, A_p)\\(b_q, B_q)\end{matrix}\right] d\theta \tag{2.8.6}$$

$$= \frac{\pi\,\Gamma(\gamma)}{2^{\gamma-1}} \exp\left(-\frac{\pi\delta}{2}\right)$$

$$\times H^{m,n}_{p+1,q+1}\left[z e^{i\pi h}\middle| \begin{matrix}(a_p, A_p), \left(\left(\dfrac{1+\gamma-i\delta}{2}\right), h\right)\\(b_q, B_q), \left(\left(\dfrac{1-\gamma-i\delta}{2}, h\right)\right)\end{matrix}\right]$$

where $h>0$, $R(\gamma)>0$, $|\arg z| < \pi\lambda/2$; $\lambda>0$ and $\mu\leqslant 0$.

$$\int_0^{\pi} (\sin\theta)^{\gamma-1} e^{-\delta\theta} H^{m,n}_{p,q}\left[z(\sin\theta)^{2l} e^{i2h\theta}\middle| \begin{matrix}(a_p, A_p)\\(b_q, B_q)\end{matrix}\right] d\theta \tag{2.8.7}$$

$$= \frac{\pi}{2^{\gamma-1}} \exp\left(-\frac{\pi\delta}{2}\right)$$

$$\times H^{m,n+1}_{p+1,q+2}\left[\frac{z e^{i\pi h}}{4^l}\middle| \begin{matrix}(1-\gamma, 2l), (a_p, A_p)\\(b_q, B_q), \left(\dfrac{1-\gamma-i\delta}{2}, l+h\right), \left(\dfrac{1-\gamma+i\delta}{2}, l-h\right)\end{matrix}\right]$$

where $h>0,\ l-h>0,\quad R(\gamma+2l\, b_j/B_j)>0\quad (j=1,\ldots,m);\ \lambda>0,$ $|\arg z|<\lambda\pi/2$ and $\mu\leqslant 0$.

When $h=1$ and $A_j=B_k=1,\ (j=1,\ldots,p,\ k=1,\ldots,q)$ then the H-function reduces to a G-function and the following results follow from (2.8.5) and (2.8.6).

$$\int_0^\pi (\sin\theta)^{\gamma-1}\, e^{-\delta\theta}\, G_{p,q}^{m,n}\left(z\sin^2\theta \,\middle|\, \begin{matrix} a_p \\ b_q \end{matrix}\right) d\theta \tag{2.8.8}$$

$$=\sqrt{\pi}\exp\left(-\frac{\pi\delta}{2}\right) G_{p+2,q+2}^{m,n+2}\left[z \,\middle|\, \begin{matrix} \dfrac{1-\gamma}{2}, \dfrac{2-\gamma}{2}, a_p \\ b_q, \dfrac{1\pm i\delta-\gamma}{2} \end{matrix}\right]$$

where $R(\gamma+2b_j)>0\ (j=1,\ldots,m);\ m+n>\dfrac{p}{2}+\dfrac{q}{2}$ and

$$|\arg z|<\left(m+n-\frac{p}{2}-\frac{q}{2}\right)\pi, \text{ and}$$

$$\int_0^\pi (\sin\theta)^{\gamma-1}\, e^{-\delta\theta}\, G_{p,q}^{m,n}\left(ze^{i2\theta} \,\middle|\, \begin{matrix} a_p \\ b_p \end{matrix}\right) d\theta \tag{2.8.9}$$

$$=\frac{\pi}{2^{\gamma-1}}\Gamma(\gamma)\, e^{-\pi\delta/2}\quad G_{p+1,q+1}^{m,n}\left[z \,\middle|\, \begin{matrix} a_p, \dfrac{1+\gamma-i\delta}{2} \\ b_q, \dfrac{1-\gamma-i\delta}{2} \end{matrix}\right]$$

where $R(\gamma)>0,\ m+n>\dfrac{p+q}{2}+1,\ |\arg z|<\left(m+n-\dfrac{p}{2}-\dfrac{q}{2}-1\right)\pi.$

2.9 AN INTEGRAL INVOLVING PRODUCT OF AN ASSOCIATED LEGENDRE FUNCTION AND AN *H*-FUNCTION

Singh and Varma (1972) have derived the following result.

$$\int_{-1}^{1} (1-x^2)^{\rho-1} P_\nu^\mu(x)\, H_{p,q}^{m,n}\left[z(1-x^2)^k \,\middle|\, \begin{matrix} (a_p, A_p) \\ (b_q, B_q) \end{matrix}\right] dx \tag{2.9.1}$$

$$=\frac{\pi 2^\mu}{\Gamma\left(\dfrac{2+\nu-\mu}{2}\right)\Gamma\left(\dfrac{1-\nu-\mu}{2}\right)}$$

$$\times H_{p+2,q+2}^{m,n+2}\left[z \,\middle|\, \begin{matrix} \left(1-\rho\pm\dfrac{\mu}{2}, k\right), (a_p, A_p) \\ (b_q, B_q), \left(1-\rho+\dfrac{\nu}{2}, k\right), \left(-\rho-\dfrac{\nu}{2}, k\right) \end{matrix}\right]$$

where $R(\rho + k\, b_j/B_j) > |R(\mu)|/2$, $(j = 1, \ldots, m)$, $k > 0$, $\lambda > 0$, $\mu \leqslant 0$ and $|\arg z| < \lambda\pi/2$.

On making use of the finite difference operator E (Milne-Thomson, 1933, p. 33 with $\omega = 1$), which has the following properties,

$$E_a f(a) = f(a+1); \tag{2.9.2}$$

$$E_a^n f(a) = E_a[E_a^{n-1} f(a)]. \tag{2.9.3}$$

Singh, F. and Varma, R.C. (1972) have further shown that

$$\int_{-1}^{1} (1-x^2)^{\rho-1} P_\nu^\mu(x)\, {}_uF_\nu(\alpha_u; \beta_\nu; c(1-x^2)^d) \times H_{p,q}^{m,n}\left[z(1-x^2)^k \,\middle|\, \begin{matrix}(a_p, A_p)\\(b_q, B_q)\end{matrix} \right] dx \tag{2.9.4}$$

$$= \frac{\pi 2^\mu}{\Gamma\left(\frac{2+\nu-\mu}{2}\right)\Gamma\left(\frac{1-\nu-\mu}{2}\right)} \sum_{r=0}^{\infty} \frac{\prod_{j=1}^{u}(\alpha_j; r)\, c^r}{\prod_{j=1}^{\nu}(\beta_j; r)\,(r)!}$$

$$\times H_{p+2,q+2}^{m,n+2}\left[z \,\middle|\, \begin{matrix}\left(1-\rho\pm\frac{\mu}{2}-rd, k\right), (a_p, A_p)\\ (b_q, B_q), \left(1-\rho-rd+\frac{\nu}{2}, k\right), \left(-\rho-rd-\frac{\nu}{2}, k\right)\end{matrix} \right],$$

where $(a \pm b, k)$ represents $(a+b, k)$ and $(a-b, k)$, $(\alpha, r) = \Gamma(\alpha+r)/\Gamma(\alpha) = \alpha(\alpha+1)\ldots(\alpha+r-1)$; $(\alpha)_0 = 1$; k and d are positive integers.

The result (2.9.4) holds if $u < \nu$ (or $u = \nu + 1$ and $|c| < 1$), no one of the (β_j) $(j = 1, \ldots, \nu)$ is zero or a negative integer, $\sum_1^p A_j - \sum_1^q B_j \leqslant 0$, $\lambda > 0$ $|\arg z| < \pi\lambda/2$ and $R(\rho + k \min b_j/B_j) > |R(\mu)|/2$ $(j = 1, \ldots, m)$.

In case $\mu = 0$ and $\nu = l$, where l is a positive integer, then (2.9.4) reduces to

$$\int_{-1}^{1} (1-x^2)^{\rho-1} P_l(x)\, {}_uF_\nu(\alpha_u; \beta_\nu; c(1-x^2)^d) \times H_{p,q}^{m,n}\left[z(1-x^2)^k \,\middle|\, \begin{matrix}(a_p, A_p)\\(b_q, B_q)\end{matrix} \right] dx \tag{2.9.5}$$

$$= \frac{\pi}{\Gamma\left(1+\frac{l}{2}\right)\Gamma\left(\frac{1-l}{2}\right)} \sum_{r=0}^{\infty} \frac{\prod_{j=1}^{u}(\alpha_j; r)\, c^r}{\prod_{j=1}^{\nu}(\beta_j; r)\,(r)!}$$

$$\times H_{p+2,q+2}^{m,n+2}\left[z \,\middle|\, \begin{matrix}(1-\rho-rd, k), (1-\rho-rd, k), (a_p, A_p)\\ (b_q, B_q), \left(1-\rho-rd+\frac{l}{2}, k\right), \left(-\rho-rd-\frac{l}{2}, k\right)\end{matrix} \right],$$

where $P_l(x)$ is the Legendre polynomial and the conditions of the validity are the same as stated in (2.9.4) with $\mu = 0$.

On the other hand, if we take $A_j = B_h = 1$ $(j = 1, \ldots, p;\ h = 1, \ldots, q)$, $k = 1$, then (2.9.4) yields

$$\int_{-1}^{1} (1-x^2)^{\rho-1} P_\nu^\mu(x)\ {}_UF_V(\alpha_U;\beta_V;\ c(1-x^2)^d)$$

$$\times G_{p,q}^{m,n}\left[z(1-x^2)\,\Big|\,\begin{matrix} a_p \\ b_q \end{matrix}\right]dx$$

$$= \frac{\pi 2^\mu}{\Gamma\left(\dfrac{2+\nu-\mu}{2}\right)\Gamma\left(\dfrac{1-\nu-\mu}{2}\right)} \sum_{r=0}^{\infty} \frac{\prod_{j=1}^{U} (\alpha_j;r)\, c^r}{\prod_{=1}^{V} (\beta_j;r)\, r!}$$

$$\times G_{p+2,q+2}^{m\cdot n+2}\left[z\,\Bigg|\,\begin{matrix} 1-\rho\pm\frac{\mu}{2}-rd,\ a_p \\ b_q,\ 1-\rho+rd+\frac{\nu}{2},\ -\rho-rd-\frac{\nu}{2} \end{matrix}\right]$$

where d is a positive integer, $U \leqslant V$ (or $U = V+1$ and $|c| < 1$);

$R(\rho + b_j) > \frac{1}{2}|R(\mu)|$, $(j = 1, \ldots, m)$; $m + n > \frac{p}{2} + \frac{q}{2}$ and

$$|\arg z| < \left(m + n - \frac{p}{2} - \frac{q}{2}\right)\pi.$$

2.10 INTEGRALS ASSOCIATED WITH GENERALIZED LAGUERRE POLYNOMIALS

The results of this section are based on the work of Shah (1969).

From the integral (Mathai and Saxena, 1973a, p. 76, 3.1.35) it can be easily shown that

$$\int_0^\infty x^\gamma e^{-x} L_k^{(\sigma)}(x)\ H_{p,q}^{m,n}\left[zx^\delta\,\Big|\,\begin{matrix}(a_p, A_p) \\ (b_q, B_q)\end{matrix}\right] dx \tag{2.10.1}$$

$$= \frac{(-1)^k (2\pi)^{(1-\delta)/2}\, \delta^{\gamma+k+1/2}}{(k)!}$$

$$\times H_{p+2\delta,q+\delta}^{m,n+2\delta}\left[z\delta^\delta\,\Big|\,\begin{matrix}(\Delta(\delta,-\gamma),1),\ (\Delta(\delta,\sigma-\gamma),1),\ (a_p, A_P) \\ (b_q, B_q),\ (\Delta(\delta,\sigma-\gamma+k),1)\end{matrix}\right]$$

where δ is a positive integer, $\mu \leqslant 0$ and $\lambda > 0$, $|\arg z| < \pi\lambda/2$, $R(\gamma + \delta b_j/B_j) > -1$ $(j = 1, \ldots, m)$.

Another interesting result for these polynomials, when δ is a positive number, has been established by Bajpai (1969) and is given below:

$$\int_0^\infty x^{\nu+\sigma/2}\, e^{-x}\, L_k^{(\sigma)}(x) \tag{2.10.2}$$

$$\times H_{p,q}^{m,n}\left[zx^{\delta} \middle| \begin{matrix}(a_p, A_p)\\(b_q, B_q)\end{matrix}\right] dx$$

$$= \frac{(-1)^k}{(k)!}\, H_{p+2,q+1}^{m,n+2}\left[z \middle| \begin{matrix}\left(-\nu - \frac{\sigma}{2}, \sigma\right), \left(-\nu + \frac{\sigma}{2}, \delta\right), \left(a_p, A_p\right)\\ \left(b_q, B_q\right), \left(-\nu + \frac{\sigma}{2} + k, \sigma\right)\end{matrix}\right]$$

The following special cases of (2.10.1) are worth mentioning and can be derived from it by specializing the parameters suitably.

$$\int_0^\infty x^{\gamma}\, e^{-x}\, L_k^{(\sigma)}(x)\, G_{p,q}^{m,n}\left(zx \middle| \begin{matrix} a_p \\ b_q \end{matrix}\right) dx \tag{2.10.3}$$

$$= \frac{(-1)^k}{k!}\, G_{p+2,q+1}^{m\ n+2}\left(z \middle| \begin{matrix} -\gamma, \sigma-\gamma, a_p \\ b_q, \sigma-\gamma+k \end{matrix}\right)$$

where $m+n > \frac{p}{2} + \frac{q}{2}$, $|\arg z| < \left(m+n-\frac{p}{2}-\frac{q}{2}\right)\pi$ and

$R(\gamma + b_j) > -1$ $(j = 1, \ldots, m)$.

$$\int_0^\infty x^{\gamma}\, e^{-x}\, L_k^{(\sigma)}(x)\, E(a_p; b_q; zx)\, dx \tag{2.10.4}$$

$$= \frac{(-1)^k}{k!}\, G_{q+3,p+1}^{p,3}\left(z \middle| \begin{matrix} -\gamma, \sigma-\gamma, 1, b_q \\ a_p, \sigma-\gamma+k \end{matrix}\right)$$

where $1-p+q \leqslant 0$, $p-q+1 > 0$, $|\arg z| < (p-q+1)\pi/2$ and $R(\gamma + a_j) > -1$ $(j = 1, \ldots, p)$ and E denotes the MacRobert's E-function.

$$\int_0^\infty x^{\gamma+l/2}\, e^{-x}\, L_k^{(\sigma)}(x)\, K_{\nu}[2(zx)^{1/2}]\, dx \tag{2.10.5}$$

$$= \frac{(-1)^k z^{-l/2}}{(2k)!}\, G_{2,3}^{2,3}\left[z \middle| \begin{matrix} -\gamma, \sigma-\gamma \\ \frac{l+\nu}{2}, \frac{l-\nu}{2}, \sigma-\gamma+k \end{matrix}\right]$$

where $|\arg z| < \pi$ and $R\left(\gamma + \frac{l}{2} \pm \frac{\nu}{2}\right) > -1$.

$$\int_0^\infty x^{\gamma+\rho} \exp\left[-x - \frac{zx}{2}\right] L_k^{(\sigma)}(x)\, W_{\lambda,\mu}(zx)\, dx \tag{2.10.6}$$

$$= \frac{(-1)^k z^{-l}}{(k)!} G^{2,2}_{3,3}\left(z \,\middle|\, \begin{matrix} -\gamma, \sigma-\gamma, \rho-\lambda+1 \\ \rho+\mu+\frac{1}{2}, \rho-\mu+\frac{1}{2}, \sigma-\gamma+k \end{matrix} \right)$$

where $|\arg z| < \pi/2$ and $R(\gamma + \rho \pm \mu) > -3/2$.

$$\int_0^\infty x^{\gamma-1/2} e^{-x} L_k^{(\sigma)}(x)\, W_{a,b}[2(zx)^{1/2}]\, W_{-a,b}[2(zx)^{1/2}]\, dx \tag{2.10.7}$$

$$= \frac{(-1)^k z^{1/2} \pi^{-1/2}}{(k)!} G^{4,2}_{4,5}\left(z \,\middle|\, \begin{matrix} -\gamma, \sigma-\gamma, \frac{1}{2}+a, \frac{1}{2}-a \\ 0, \frac{1}{2}, b, -b, \sigma-\gamma+k \end{matrix} \right)$$

where $|\arg z| < \pi/2$ and $R(\gamma \pm b) > -1$.

$$\int_0^\infty x^{\gamma} e^{-x} L_k^{(\sigma)}(x)\, I_a[(zx)^{1/2}]\, K_a[(zx)^{1/2}]\, dx \tag{2.10.8}$$

$$= \frac{(-1)^k \pi^{-1/2}}{(2k)!} G^{2,3}_{3,4}\left[z \,\middle|\, \begin{matrix} -\gamma, \sigma-\gamma, \frac{1}{2} \\ a, 0, -a, \sigma-\gamma+k \end{matrix} \right]$$

where $|\arg z| < \pi$ and $R(\gamma \pm a) > -1$.

$$\int_0^\infty x^{\gamma} e^{-x} L_k^{(\sigma)}(x)\, {}_p\psi_q\left[\begin{matrix}(a_p, A_p) \\ (b_q, B_q)\end{matrix}; -zx\right] dx \tag{2.10.9}$$

$$= \frac{(-1)^k}{k!} H^{1,p+2}_{p+2,q+2}\left[z \,\middle|\, \begin{matrix} (-\gamma, 1), (\sigma-\gamma, 1), (1-a_p, A_p) \\ (0, 1), (1-b_q, B_q), (\sigma-\gamma+k, 1) \end{matrix} \right]$$

where $-1 < \mu \leqslant 1$, $|\arg z| < \pi(1+\mu)/2$ and $R(\gamma) > -1$.

$$\int_0^\infty x^{\gamma} e^{-x} L_k^{(\sigma)}(x)\, J_\nu^\mu(zx)\, dx \tag{2.10.10}$$

$$= \frac{(-1)^k}{(k)!} H^{1,2}_{2,3}\left[z \,\middle|\, \begin{matrix} (-\gamma, 1), (\sigma-\gamma, 1) \\ (0, 1), (-\nu, \mu), (\sigma-\gamma+k, 1) \end{matrix} \right]$$

where $R(\gamma) > -1$ and $|\arg z| < \pi/2$.

EXERCISES

2.1 Prove that

$$\int_0^\infty x^{-u} J_\omega(x) J_\nu(x) H^{0,N,N_1,M,M_1}_{E,(A:C),F,(B:D)}\left[\begin{matrix} ax^{2h} \\ y \end{matrix}\right] dx$$

$$= 2^{-u} H^{0,N+1,N_1,M+1,M_1}_{E,(A+4:C),F,(B+1:D)}\left[\begin{matrix} 2^{2h}a \\ y \end{matrix}\middle| \chi_1\right]$$

where χ_1 stands for the parameters

$$\chi_1 = \left[\begin{matrix} (e_E, \theta_E) \\ (\rho, h), (a_A, \alpha_A), (\xi, h), (\eta, h) \\ (\zeta, h) \; ; \; (c_C, \gamma_C) \\ (f_F, \phi_F) \\ (u, 2h), (b_B, \beta_B) \; ; \; (d_D, \delta_D) \end{matrix}\right]$$

where

$$\rho = \frac{1+u-\nu-\omega}{2}, \quad \xi = \frac{1+u+\nu+\omega}{2},$$

$$\eta = \frac{1+u+\nu-\omega}{2}, \quad \zeta = \frac{1+u+\omega-\nu}{2},$$

$$|\arg a| < \rho_1 \pi/2$$

and

$$R(\omega + \nu - u + 2h\, b_j/B_j) > -1, \; (j = 1, \ldots, m),$$

$$R(2ha_j/\alpha_j - u) < 2h/\alpha_j \; (j = 1, \ldots, n);$$

and

$$\rho_1 = \sum_1^M \beta_j - \sum_{M+1}^B \beta_j + \sum_1^N \alpha_j - \sum_{N+1}^A \alpha_j - \sum_1^E \theta_j - \sum_1^F \phi_j > 0$$

(Bora, S.L., Saxena, R.K. and Kalla, S.L., 1972)

2.2 Establish the formula

$$\int_0^\infty t^{\rho-1} W_{k,\mu}(t)\, W_{\lambda\nu}(t)\, H\left[\begin{matrix} xt^\sigma \\ yt^\sigma \end{matrix}\right] dt$$

$$= \sum_{\nu,-\nu} \frac{\Gamma(-2\nu)}{\Gamma(\frac{1}{2} - \lambda - \nu)} \sum_{r=0}^\infty \frac{(\frac{1}{2} - \lambda + \nu)_r}{r!\,(1+2\nu)_r}$$

$$\times H^{L+2,N,N_1,M,M_1}_{E+2,(A:C),F+1,(B:D)}\left[\begin{array}{c|c} x & (e_E,\theta_E),(1\pm\mu+\rho+\nu+r,\sigma) \\ & (a_A,\alpha_A);(c_C,\gamma_C) \\ y & (f_F,\phi_F);(3/2-k+\nu+\rho+r,\sigma) \\ & (b_B,\beta_B);\ (d_D,\delta_D) \end{array}\right]$$

where σ is a positive number,

$$R(\rho+\sigma\min(b_j/\beta_j)+\sigma\min(d_h/\delta_h)] > |R(\mu)|+|R(\nu)|-1$$

for $j=1,\ldots,M$ and $h=1,\ldots,M_1$; $\rho_j>0\ (j=1,2)$; $|\arg x|<\pi\rho_1/2$, $|\arg y|<\pi\rho_2/2$ and ρ_j are defined in Section 2.2.

Hence or otherwise obtain the value of the following integral in the form given below:

$$\int_0^\infty t^{\rho-1/2}e^{-t/2}W_{k,\mu}(t)\,H\begin{bmatrix} xt^\sigma \\ yt^\sigma \end{bmatrix}dt$$

$$= H^{L+2,N,N_1,M,M_1}_{E+2,(A:C),F+1,(B:D)}\left[\begin{array}{c|c} x & (e_E,\theta_E),(1\pm\mu+\rho,\sigma) \\ & (a_A,\alpha_A);(c_C,\gamma_C) \\ y & (f_F,\phi_F),(3/2-k+\rho,\sigma) \\ & (b_B,\beta_B);(d_D,\delta_D) \end{array}\right]$$

(Saxena, 1971, pp. 5–6)

2.3 Show that

$$a\int_0^\infty H^{P,O}_{Q,R}\left[ax\,\middle|\,\begin{matrix}(q_Q,\psi_Q)\\(-r_R,\zeta_R)\end{matrix}\right]\ H\begin{bmatrix} bx \\ cx \end{bmatrix}dx$$

$$= H^{L+P,N,N_1,M,M_1}_{E+Q,[A:C],F+R,[B:D]}\left[\begin{array}{c|c} \dfrac{b}{a} & (e_E,\theta_E);(\zeta_R-r_R,\zeta_R) \\ & (a_A,\alpha_A);(c_C,\gamma_C) \\ \dfrac{c}{a} & (f_F,\phi_F);(q_Q+\psi_Q-1,\psi_Q) \\ & (b_B,\beta_B):(d_D,\delta_D) \end{array}\right]$$

(Saxena, 1971, p. 190, 5.1)

2.4 Establish the following result

$$\int_0^\infty t^{\rho-1}\,H^{M,N}_{L,U}\left[vt\,\middle|\,\begin{matrix}(a'_L,A'_L)\\(b'_U,B'_U)\end{matrix}\right]$$

$$\times\ H^{L,N,N_1,M,M_1}_{E,(A:C),F,(B:D)}\left[\begin{matrix}\sigma^{\sigma N}\ xt^{\sigma}\\ \sigma^{\sigma N}\ yt^{\sigma}\end{matrix}\left|\begin{matrix}(e_E,\theta_E),\ \triangle(\sigma;1-a'_N-\rho A'_N)\\ (a_A,\alpha_A);(c_C,\gamma_C)\\ (f_F,\phi_F)\\ (b_B,\beta_B)\ ;\ (d_D,\delta_D)\end{matrix}\right.\right]dt$$

$$=\nu^{-\rho}\ H^{L+M,N,N_1,M,M_1}_{E+U,(A:C),F+L-N,(B:D)}\left[\begin{matrix}\nu^{\sigma}\ \sigma^{\sigma N}x\\ \nu^{\sigma}\ \sigma^{\sigma N}y\end{matrix}\middle|\ \chi\right]$$

where χ represents the parameters,

$$\left[\begin{matrix}\triangle_{\sigma}(-\sigma,b'_M+\rho B'_M),(e_E,\theta_E),\triangle(\sigma:1-b'_{M+1,u}-\rho B'_{M+1,u})\\ (a_A,\alpha_A)\ ;\ (c_C,\gamma_C)\\ (f_F,\phi_F)\ ;\ \triangle(\sigma,a'_{N+1,L}+\rho A'_{N+1,L})\\ (b_B,\beta_B)\ :\ (d_D,\delta_D)\end{matrix}\right]$$

where $\triangle(\sigma,a_{M+1,U})$ represents the $U-M$ parameters

$$\triangle(\sigma,a_{M+1}),\ \triangle(\sigma,a_{M+2}),\ldots,\ \triangle(\sigma,a_U).$$

(Banerji and Saxena, 1973)

2.5 Evaluate the following integrals

(i)
$$\prod_{r=1}^{t}\int_0^1 x_r^{\alpha_r-1}(1-x_r)^{-1/2}\,T_{n_r}(2x_r-1)$$
$$\times\ H^{m,n}_{p,q}\left[\frac{z}{(x_1x_2,\ldots,x_t)^h}\left|\begin{matrix}(a_p,A_p)\\(b_q,B_q)\end{matrix}\right.\right]dx_r$$
$$=\sqrt{\pi}\ H^{m+2t,n}_{p+2t,q+2t}\left[z\left|\begin{matrix}(a_p,A_p),(a_1-n_1+\tfrac12,h),(a_1+n_1+\tfrac12,h),\\ \ldots,(a_t-n_t+\tfrac12,h),(a_t+n_t+\tfrac12,h)\\ (b_q,B_q),(\alpha_1,h),(\alpha_1+\tfrac12,h),\ldots,\\ (a_t,h),(\alpha_t+\tfrac12,h)\end{matrix}\right.\right]$$

where $R(h\,a_j/A_j-\alpha_r)<h/A_j\ (j=1,\ldots,n;\ r=1,\ldots,t),\ \mu\leqslant 0,\ \lambda>0,\ h>0$ and $|\arg z|<\lambda\pi/2$.

(ii)
$$\prod_{r=1}^{t}\int_0^1 x_r^{\alpha_r-1}(1-x_r)^{-1/2}\,T_{n_r}(2x_r-1)$$
$$\times\ H^{m,n}_{p,q}\left[z(x_1x_2\ldots x_t)^h\left|\begin{matrix}(a_p,A_p)\\(b_q,B_q)\end{matrix}\right.\right]dx_r$$
$$=\sqrt{\pi}\ H^{m,n+2t}_{p+2t,q+2t}\left[z\left|\begin{matrix}(1-\alpha_1,h),(\tfrac12-\alpha_1,h),\ldots,(1-\alpha_t,h),\\ (\tfrac12-\alpha_t,h),(a_p,A_p)\\ (b_q,B_q),(\tfrac12-\alpha_1-n_1,h),(\tfrac12-a_1+n_1,h),\\ \ldots,(\tfrac12-\alpha_t-n_t,h),(\tfrac12-\alpha_t+n_t,h)\end{matrix}\right.\right]$$

where $h > 0$, $\mu \leqslant 0$, $\lambda \geqslant 0$, $|\arg z| < \lambda\pi/2$, $R(\alpha_r + h\, b_j/B_j) > -1$, $(j = 1, \ldots, m)$.

2.6 Prove that

$$\int_{-1}^{1} (1+t)^{\mu-1}(1-t)^{\lambda-1} P_\nu^{(\alpha,\beta)}\left[1 - \frac{\gamma z}{2}(1-t)\right] H\begin{bmatrix} x(1-t)^r \\ y(1-t)^r \end{bmatrix} dt$$

$$= \frac{2^{\lambda+\mu-1}(1+\alpha)_\nu \Gamma(\mu)}{(\nu)!} \sum_{k=0}^{\nu} \frac{(-\nu)_k (1+\alpha+\beta+\nu)_k \left(\dfrac{\gamma z}{2}\right)^k}{k!\,(1+\alpha)_k}$$

$$\times H^{L+1,N,N_1,M,M_1}_{E+1,(A:C),F+1,(B:D)}\left[\begin{matrix} 2^r x \\ 2^r y \end{matrix}\middle|\, \chi_1\right]$$

where χ_1 denotes the parameters

$$\begin{bmatrix} (\lambda+k, r), (e_E, \theta_E) \\ (a_A, \alpha_A);\ (c_C, \alpha_C) \\ (\lambda+\mu+k, r) \\ (b_B, \beta_B);\ (d_D, \delta_D) \end{bmatrix}$$

where $r > 0$, $\rho_j > 0$, $(j = 1, 2)$; $|\arg x| < \pi\rho_1/2$, $|\arg y| < \frac{1}{2}\pi\rho_2$, $R(\mu) > 0$, $R\left(\lambda + r\dfrac{b_j}{\beta_j} + r\dfrac{d_h}{\delta_h}\right) > 0$ $(j = 1, \ldots, M;\ h = 1, \ldots, M_1)$ and $P_\nu^{(\alpha,\beta)}(t)$ is the Jacobi polynomial.

When $L = E = F = 0$, deduce that

$$\int_{-1}^{1} (1+t)^{\mu-1}(1-t)^{\lambda-1} P_\nu^{(\alpha,\beta)}\left\{1 - \frac{\gamma z}{2}(1-t)\right\}$$

$$\times H^{M,N}_{A,B}\left[x(1-t)^r \middle| \begin{matrix} (a_A, \alpha_A) \\ (b_B, \beta_B) \end{matrix}\right] H^{M_1,N_1}_{C,D}\left[y(1-t)^r \middle| \begin{matrix} (c_C, \gamma_C) \\ (d_D, \delta_D) \end{matrix}\right] dt$$

$$= \frac{2^{\lambda+\mu-1}(1+\alpha)_\nu \Gamma(\mu)}{(\nu)!} \sum_{k=0}^{\nu} \frac{(-\nu)_k (1+\alpha+\beta+\nu)_k \left(\dfrac{\gamma z}{2}\right)^k}{k!\,(1+\alpha)_k}$$

$$\times H^{1,N,N_1,M,M_1}_{1,[A:C],1,[B:D]}\left[\begin{matrix} 2^r x \\ \\ 2^r y \end{matrix}\middle| \begin{matrix} (\lambda+k, r) \\ (a_A, \alpha_A);\ (c_C, \gamma_C) \\ (\lambda+\mu+k, r) \\ (b_B, \beta_B),\ (d_D, \delta_D) \end{matrix}\right]$$

(Munot and Kalla, 1971)

Finally show that

$$\int_{-1}^{1} (1+t)^{\mu-1}(1-t)^{\lambda-1}\, P_\nu^{(\alpha,\beta)}\left[1-\frac{\gamma y}{2}(1-t)\right]$$

$$\times H_{p,q}^{m,n}\left[z(1-t)^h \,\middle|\, \begin{matrix}(a_p, A_p)\\(b_q, B_q)\end{matrix}\right] dt$$

$$= \frac{2^{\lambda+\mu-1}(\alpha+1)_\nu\,\Gamma(\mu)}{\nu!} \sum_{r=0}^{\nu} \frac{(-\nu)_r}{r!}\,\frac{(1+\alpha+\beta+\nu)_r}{(1+\alpha)_r}\left(\frac{\gamma y}{2}\right)^r$$

$$\times H_{p+1,q+1}^{m,n+1}\left[2^h z \,\middle|\, \begin{matrix}(1-\lambda-r, h), (a_p, A_p)\\(b_q, B_q), (1-\lambda-\mu-r, h)\end{matrix}\right]$$

$$h>0,\quad \sum_1^p A_j - \sum_1^q B_j \leqslant 0,\quad \sum_1^n A_j - \sum_{n+1}^p A_j + \sum_1^m B_j - \sum_{m+1}^q B_j > 0$$

$$|\arg z| < \tfrac{1}{2}\pi\left(\sum_1^n A_j - \sum_{n+1}^p A_j + \sum_1^m B_j - \sum_{m+1}^q B_j\right),$$

$$R(\mu) > 0,\quad R\left(\lambda + h\frac{b_j}{B_j}\right) > 0,\quad (j=1, \ldots, m).$$

(Bajpai, 1969a)

2.7 Prove that

$$\int_0^\infty \frac{1}{\Gamma(\pm 2it)}\, G_{2,3}^{3,1}\left[x \,\middle|\, \begin{matrix}1, a\\ b, \frac{1}{2}\pm it\end{matrix}\right] H_{p,\,q+2}^{m+2,\,n}\left[y \,\middle|\, \begin{matrix}(a_p, A_p)\\(c\pm it, k), (b_q, B_q)\end{matrix}\right] dt$$

$$= 2\pi\, x^{c+1/2}\, H_{p+1,q+2}^{m+1,n+1}\left[\frac{y}{x^k} \,\middle|\, \begin{matrix}(\frac{3}{2}-b-c, k)\ (a_p, A_p)\\ (\frac{1}{2}+c, k), (b_q, B_q), (\frac{3}{2}-a+c, k)\end{matrix}\right]$$

where $R(x)>0$, $R(b)>0$, $\lambda>0$, $|\arg y| < \pi\lambda/2$, $R(a-c+k\, b_j/B_j) > \frac{1}{2}$,

$$R(b-c+k\, b_j/B_j) > 0 \text{ and } R\left[c - \frac{k}{A_h}(a_h-1)\right] > 0$$

$$(j=1,\ldots,m;\ h=1, \ldots, n).$$

Hence deduce the following results.

(i) ${}_2F_1(a, b, c; -y/x)$

$$= \frac{e^{y/2}\,\Gamma(c)}{2\pi y\, \Gamma(a)\,\Gamma(b)} \int_0^\infty \frac{\Gamma(b\pm it-\frac{1}{2})}{\Gamma(\pm 2it)}\, W_{1-b,\,it}(y)$$

$$\times G_{2,3}^{3,1}\left[x \,\middle|\, \begin{matrix}1, c\\ a, \frac{1}{2}\pm it\end{matrix}\right] dt$$

where $R(x)>0$, $R(a)>0$, $R(b)>\frac{1}{2}$ and $|y/x|<1$.

(ii) $$W_{k,m}(x/y) = \frac{e^{x/2y}\, y^{m-1/2}}{2\pi\, x^{m+l}\, \Gamma(l+\frac{1}{2})}$$

$$\times \int_0^\infty {}_2F_1\left(l \pm it; l+\frac{1}{2}; -\frac{1}{y}\right) \frac{\Gamma(l \pm it)}{\Gamma(\pm 2it)}$$

$$\times G_{2,3}^{3,1}\left[x \,\middle|\, \begin{matrix} 1, 1+l-k+m \\ \frac{1}{2}+l+2m, \frac{1}{2} \pm it \end{matrix}\right] dt,$$

where $R(x) > 0$, $|y| > 1$, $R(l) > 0$ and $R(\frac{1}{2} + l + 2m) > 0$.

(Samar, 1973)

2.8 Establish the following results.

$$\int_0^\infty t^{\lambda-1} J_\nu(pt)\, H^{(1)}\begin{bmatrix} yt^\delta \\ zt^\delta \end{bmatrix} dt$$

$$= \frac{2^{\lambda-1}}{p^\lambda} H^{1,N,N_1,M,M_1}_{E+2,[A:C],F,[B:D]} \left[\begin{matrix} \left(\frac{2}{p}\right)^\delta y \\ \left(\frac{2}{p}\right)^\delta z \end{matrix} \,\middle|\, x \right],$$

where χ represents the parameters

$$\left[\begin{matrix} \left(\frac{\lambda}{2} \pm \frac{\nu}{2}, \frac{\delta}{2}\right), (e_E, \theta_E) \\ (a_A, \alpha_A) \;;\; (c_C, \gamma_C) \\ (f_F, \phi_F) \\ (b_B, \beta_B) \;;\; (d_D, \delta_D) \end{matrix}\right]$$

where $R\left(\lambda + \nu + \delta \min \frac{b_j}{\beta_j} + \delta \min \frac{d_k}{\delta_k}\right) > 0 \; (j = 1, \ldots, M; k = 1, \ldots, M_1)$,

$R\left[\lambda + \delta \max\left(\frac{a_j - 1}{A_j}\right) + \delta \max\left(\frac{c_h - 1}{\gamma_h}\right)\right] < \frac{3}{2}, (j = 1, \ldots, N; h = 1, \ldots, N_1)$;

$p > 0$, $\delta > 0$, $|\arg y| < \frac{\pi\rho_1}{2}$, $|\arg z| < \frac{\pi\rho_2}{2}$, $\rho_1 > 0$, $\rho_2 > 0$.

(Munot, 1972)

Hence derive the following integrals.

(i) $$\int_0^\infty t^{\sigma+1/2} J_\nu(st)\, H^{m,n}_{p,q}\left[zt^\delta \,\middle|\, \begin{matrix} (a_p, A_p) \\ (b_q, B_q) \end{matrix}\right] dt = 2^{\sigma+1/2} s^{-\sigma-3/2}$$

$$\times H^{m,n+1}_{p+2,q}\left[\left(\frac{2}{s}\right)^\delta z \,\middle|\, \begin{matrix} \left(\frac{1}{4} - \frac{\sigma}{2} - \frac{\nu}{2}, \frac{\delta}{2}\right), (a_p, A_p), \left(\frac{1}{4} - \frac{\sigma}{2} + \frac{\nu}{2}, \frac{\delta}{2}\right) \\ (b_q, B_q) \end{matrix}\right],$$

where $\delta > 0$, $s > 0$, $R[\sigma + \nu + \delta \min (b_j/B_j)] > -\frac{3}{2}$, $(j = 1, \ldots, m)$; $\sum_1^p A_j - \sum_1^q B_j < 0$, $R\left[\sigma + \delta \max\left(\frac{a_j - 1}{A_j}\right)\right] < 0$ $(j = 1, \ldots, n)$; $\lambda > 0$, $|\arg z| < \frac{\pi\lambda}{2}$.

(ii) $$\int_0^\infty t^{\lambda-1} J_\nu(pt)\, H_{A,B}^{M,N}\left[yt^\delta \middle| \begin{matrix}(a_A, \alpha_A)\\(b_B, \beta_B)\end{matrix}\right] H_{C,D}^{M_1,N_1}\left[zt^\delta \middle| \begin{matrix}(c_C, \gamma_C)\\(d_D, \delta_D)\end{matrix}\right] dt$$

$$= \frac{2^{\lambda-1}}{p^\lambda} H^{1,N,N_1,M,M_1}_{2,[A:C],0,[B:D]}\left[\begin{matrix}\left(\frac{2}{p}\right)^\delta y\\ \left(\frac{2}{p}\right)^\delta z\end{matrix} \middle| \chi_1\right],$$

where χ_1 denotes the parameters

$$\left[\begin{matrix} & \left(\frac{\lambda \pm \nu}{2}, \frac{\delta}{2}\right) & \\ (a_A, \alpha_A) & ; & (c_C, \gamma_C) \\ (b_B, \beta_B) & ; & (d_D, \delta_D)\end{matrix}\right]$$

and the conditions of validity are as follows.

$R\left[\lambda + \delta \min\left(\frac{b_j}{B_j}\right) + \delta \min\left(\frac{d_h}{\delta_h}\right) + \nu\right] > 0$, $(j = 1, \ldots, M,\ h = 1, \ldots, M_1)$,

$R\left[\lambda + \delta \max\left(\frac{a_j - 1}{\alpha_j}\right) + \delta \max\left(\frac{c_h - 1}{\gamma_h}\right)\right] < \frac{3}{2}$, $(j=1,\ldots,N;\ h=1,\ldots,N_1)$,

$p > 0$, $\sum_1^A \alpha_j - \sum_1^B \beta_j \leqslant 0$; $\sum_1^C c_j - \sum_1^D d_j \leqslant 0$, $|\arg y| < \frac{\pi\lambda_1}{2}$ and $|\arg z| < \frac{\pi\lambda_2}{2}$

where λ_1 and λ_2 are defined in Section 2.5.

(iii) $$\int_0^\infty t^{-\rho} J_\nu(t) J_\omega(t)\, H_{p,q}^{m,n}\left[zt^{2h}\middle|\begin{matrix}(a_p, A_p)\\(b_q, B_q)\end{matrix}\right] dt$$

$$= 2^{-\rho} H^{m+1,n+1}_{p+4,q+1}\left[2^{2h} z \middle| \begin{matrix}\left(\frac{1+\rho-\omega-\nu}{2}, h\right)(a_p, A_p), \left(\frac{1+\rho+\nu\pm\omega}{2}, h\right)\\ \left(\frac{\rho+\omega-\nu+1}{2}, h\right), (\rho, 2h), (b_q, B_q)\end{matrix}\right]$$

where $h > 0$, $\mu \leqslant 0$, $|\arg z| < \frac{\pi\lambda}{2}$, $\lambda > 0$

$$R[\omega+\nu-\rho+2h \min b_j/B_j] > -1\ (j = 1, \ldots, m)$$

and $\quad R\,[2h\,(a_j/A_j) - \rho] < \dfrac{2h}{A_j}.$

2.9 Show that

$$\frac{1}{2\pi i}\int_{\sigma-i\infty}^{\sigma+i\infty} e^{\omega t}\,(\omega+\rho)^{-\mu}\, S_n^m\left(\frac{\lambda}{\omega+\rho}\right) H_{P,Q}^{M,N}\left[z\,(\omega+\rho)^{\nu}\,\middle|\,\begin{matrix}(a_p, A_p)\\(b_q, B_q)\end{matrix}\right] d\omega$$

$$= t^{\mu-1}\, e^{-t\rho} \sum_{k=0}^{[n/m]} \frac{(-n)_{mk}\, A_{n,k}\,(\lambda t)^k}{k!}$$

$$\times\, H_{P+1,Q}^{M,N}\left[\frac{z}{t^{\nu}}\,\middle|\,\begin{matrix}(a_P, A_P), (\mu+k, \nu)\\(b_Q, B_Q)\end{matrix}\right]$$

where $\quad \nu > 0,\ R(\mu) > 0,\ \ A_j > 0\ (j = 1, \ldots, P),\ \ B_j > 0\ (j = 1, \ldots, Q);$
$|\arg z| < \pi R^*/2,\ \sum_1^P A_j - \sum_1^Q B_j \leqslant 0$

$$R^* = \sum_1^N A_j - \sum_{N+1}^P A_j + \sum_1^M B_j - \sum_{M+1}^Q B_j > 0$$

and

$$S_n^m\,(x) = \sum_{k=0}^{[n/m]} \frac{(-n)_{mk}}{k!}\, A_{n,k}\, x^k,\ n = 0, 1, 2, \ldots;$$

where m is an arbitrary positive integer and the $A_{n,k}$ are arbitrary constants real or complex. (Srivastava, H.M., 1972, p. 2)

2.10 Prove the following integrals.

(i) $$\int_0^{\infty} t^{\lambda-1}\, K_{\nu}(at)\, H_{A,B}^{M,N}\left[bt^{2h}\,\middle|\,\begin{matrix}(a_A, \alpha_A)\\(b_B, \beta_B)\end{matrix}\right] H_{C,D}^{M_1,N_1}\left[ct^{2h}\,\middle|\,\begin{matrix}(c_C, \gamma_C)\\(d_D, \delta_D)\end{matrix}\right] dt$$

$$= 2^{\lambda-2}\, a^{-\lambda}\, H_{2,(A:C),0,(B:D)}^{2,N,N_1,M,M_1}\left[\begin{matrix} b\left(\dfrac{2}{a}\right)^h \\ c\left(\dfrac{2}{a}\right)^h\end{matrix}\,\middle|\,\begin{matrix}\left(\dfrac{\lambda+\nu}{2}, h\right), \left(\dfrac{\lambda-\nu}{2}, h\right)\\ a_A, \alpha_A);\ (c_C, \gamma_C)\\ (b_B, \beta_B);\ (d_D, \delta_D)\end{matrix}\right]$$

where

$$R\left[\lambda \pm \nu + 2h\,\min\left(\frac{b_j}{\beta_j}\right) + 2h\,\min\left(\frac{d_h}{\delta_h}\right)\right] > 0\,(j=1,\ldots,M;\ h=1,\ldots,M_1);$$

$$R(a) > 0,\ \sum_1^A \alpha_j - \sum_1^B \beta_j \leqslant 0,\ \sum_1^C \gamma_j - \sum_1^D \delta_j \leqslant 0,$$

$|\arg b| < \pi\lambda_1/2$, $|\arg c| < \pi\lambda_2/2$, where λ_1 and λ_2 are defined in (2.5.1).

(ii) $$\int_0^\infty t^{\lambda-1} e^{-at} K_\nu(at) H_{A,B}^{M,N}\left[bt^h \middle| \begin{matrix}(a_A, \alpha_A)\\(b_B, \beta_B)\end{matrix}\right] H_{C,D}^{M_1,N_1}\left[ct^h \middle| \begin{matrix}(c_C, \gamma_C)\\(d_D, \delta_D)\end{matrix}\right] dt$$

$$=\frac{\Gamma(\frac{1}{2})}{2^\lambda a^\lambda} H_{2,(A:C),1,(B:D)}^{2,N,N_1,M,M_1}\left[\begin{matrix}\dfrac{b}{(2a)^h}\\ \dfrac{c}{(2a)^h}\end{matrix} \middle| \begin{matrix}(\lambda\pm\nu, h)\\(a_A, \alpha_A); (c_C, \gamma_C)\\(\lambda+\frac{1}{2}, h)\\(b_B, \beta_B); (d_D, \delta_D)\end{matrix}\right],$$

where

$$R\left[\lambda\pm\nu+h\min\left(\frac{b_j}{\beta_j}\right)+h\min\left(\frac{d_h}{\delta_h}\right)\right]>0, \ (j=1,\ldots,M;\ h=1,\ldots,M_1);$$

$$R(a)>0, \ \sum_1^A \alpha_j - \sum_1^B \beta_j \leqslant 0, \sum_1^C \gamma_j - \sum_1^D \delta_j \leqslant 0_1,$$

$$|\arg b| < \frac{\pi\lambda_1}{2} \ \text{ and } \ |\arg c| < \frac{\pi\lambda_2}{2}.$$

(iii) $$\int_0^\infty t^{\sigma-1} K_\nu(at) H_{p,q}^{m,n}\left[zt^\delta \middle| \begin{matrix}(a_p, A_p)\\(b_q, B_q)\end{matrix}\right] dt$$

$$=2^{\sigma-2} a^{-\sigma} H_{p+2,q}^{m,n+2}\left[\left(\frac{2}{a}\right)^\delta z \middle| \begin{matrix}\left(1-\dfrac{\sigma}{2}\pm\dfrac{\nu}{2}, \dfrac{\delta}{2}\right), (a_p, A_p)\\(b_q, B_q)\end{matrix}\right]$$

where

$R[\sigma\pm\nu+\delta \min(b_j/B_j)]>0 \ (j=1,\ldots,m)$, $R(a)>0$; $\lambda>0$, $\mu\leqslant 0$ and $|\arg z|<\pi\lambda/2$.

(iv) $$\int_0^\infty t^{\rho-1} K_\mu(at) K_\nu(at) H_{A,B}^{M,N}\left[bt^{2h} \middle| \begin{matrix}(a_A, \alpha_A)\\(b_B, \beta_B)\end{matrix}\right]$$

$$\times H_{C,D}^{M_1,N_1}\left[ct^{2h} \middle| \begin{matrix}(c_C, \gamma_C)\\(d_D, \delta_D)\end{matrix}\right] dt$$

$$=2^{\rho-3} a^{-\rho} H_{4,(A:C),1,(B:D)}^{4,N,N_1,M,M_1}\left[\begin{matrix}b\left(\dfrac{2}{a}\right)^{2h}\\ c\left(\dfrac{2}{a}\right)^{2h}\end{matrix} \middle| \begin{matrix}\left(\dfrac{\rho\pm\mu\pm\nu}{2}, h\right)\\(a_A, \alpha_A); (c_C, \gamma_C)\\(\rho, 2h)\\(b_B, \beta_B); (d_D, \delta_D)\end{matrix}\right]$$

where

$$R(a)>0, \ R[\rho\pm\mu\pm\nu+2h\min(b_j/B_j)+2h\min(d_h/\delta_h)]>0,$$

$$(j=1,\ldots,M;\ h=1,\ldots,M_1);\ \lambda_1,\ \lambda_2>0;$$

$$|\arg b|<\pi\lambda_1/2 \text{ and } |\arg c|<\pi\lambda_2/2;$$

$$\sum_1^A \alpha_j - \sum_1^B \beta_j \leqslant 0, \ \sum_1^C \gamma_j - \sum_1^D \delta_j \leqslant 0,$$

$|\arg b| < \pi\lambda_1/2$ and $|\arg c| < \pi\lambda_2/2$, where λ_1 and λ_2 are defined in (2.5.1).

2.11 Prove that

$$\int_0^\infty x^{\lambda-1} G_{p,q}^{m,n}\left[zx \middle| \begin{matrix}\sigma_1, \ldots, \sigma_p \\ \tau_1, \ldots, \tau_q\end{matrix}\right] H\left[\begin{matrix} U \\ Vx^r \end{matrix}\right] dx$$

$$= z^{-\lambda} H^{L,N,N_1+n,M_1+m}_{E,[A:C+q-m],F,(B\dagger D+p-n)}\left[\begin{matrix} U \\ V/z^r \end{matrix} \middle| \chi \right]$$

where χ stands for the parameters

$$\left[\begin{matrix} (e_E, \theta_E) \\ (a_A, \alpha_A); k_1, \ldots, k_m; (c_C, \gamma_C), k_{m+1}, \ldots, k_q \\ (f_F, \phi_F) \\ (b_B, \beta_B);\ \ l_1, \ldots, l_n,\ \ (d_D, \delta_D),\ \ l_{n+1}, \ldots, l_p \end{matrix}\right]$$

$k_j = (1 - \lambda - \tau_j, r)$; $j = 1, \ldots, q$; $l_j = (1 - \lambda - \sigma_j, r)$; $j = 1, \ldots, p$; $|\arg U| < \pi\lambda_1/2$ and $|\arg V| < \pi\lambda_2/2$; $R[\lambda + \tau_j + r(d_h/\delta_h)] > 0$ $(j = 1, \ldots, m;\ h = 1, \ldots, M_1)$; $R[\lambda - \sigma_i + r(c_j - 1)/\gamma_j - rl_h] < -1$ $(i = 1, \ldots, n;\ j = 1, \ldots, N_1$ and $h = 1, \ldots, L)$,

$$|\arg z| < \left(m + n - \frac{p}{2} - \frac{q}{2}\right)\pi \text{ and } m + n > \frac{p}{q} + \frac{q}{2}.$$

(Kalla, 1972)

REMARK. For the integration of H-function of two variables with respect to a parameter, see the work of de Anguio and Kalla (1973).

2.12 Prove the following results.

$$\int_0^1 t^{\alpha-1}(1-t)^{\beta-1}\ {}_2F_1\left(\nu, -k; \varepsilon; \frac{t}{t-1}\right) H_{p,q}^{m,n}\left[\frac{z(1-t)^l}{t^l} \middle| \begin{matrix}(a_p, A_p) \\ (b_q, B_q)\end{matrix}\right] dt$$

$$= \frac{(2\pi)^{1-l}\ l^{2\alpha+2\beta+\nu-\varepsilon-k-1}}{(\varepsilon)_n\ \Gamma(\varepsilon - \nu)}$$

$$\times H^{l+m,l+n}_{2l+p,2l+q}\left[z \middle| \begin{matrix}\Delta_l(-l; \varepsilon - \alpha - l - k), (a_p, A_p), \Delta_l(-l, \beta - k - l) \\ \Delta(l, \alpha), (b_q, B_q), \Delta_l(-l; \varepsilon - \alpha - k)\end{matrix}\right],$$

where $\varepsilon + k = \alpha + \beta + \nu$; $\sum_1^p A_j - \sum_1^q B_j \leqslant 0$;

$$R[\alpha - l(a_j - 1)/A_j] > 0, (j = 1, \ldots, n);$$

$$R\left[\beta + \nu - l\frac{b_j}{B_j}\right] > 0 \quad (j = 1, \ldots, m);$$

and

$$R\left[\beta - k - l\frac{b_j}{B_j}\right] > 0 \quad (j = 1, \ldots, m)$$

$\lambda > 0$, $|\arg z| < \pi\lambda/2$ and the symbol $\Delta_m(p, a)$ indicates m parameters of the type

$$\left[\frac{a}{p}, \frac{a+1}{p}, \ldots, \frac{a+m-1}{p}\right]$$

and $\Delta(n, b)$ represents n parameters of the type b/n, $(b+1)/n, \ldots,$ $(b+n-1)/n$. The parameters λ and μ are defined in (2.6.6) and (2.6.7).

$$\int_0^1 t^{2\alpha-1}(1-t)^{2\beta-1}\, {}_2F_1(\alpha+\beta, \nu; 2\nu; 4t(1-t))$$

$$\times H_{p,q}^{m,n}\left[\frac{zt^{2l}}{(1-t)^{2l}}\middle| \begin{matrix}(a_p, A_p)\\(b_q, B_q)\end{matrix}\right] dt$$

$$= \tfrac{1}{2} l^{2\alpha+2\beta-2\nu-1} \frac{\Gamma(\nu/2)\,\Gamma(\frac{1}{2}-\alpha-\beta+\nu)}{\Gamma(\alpha+\beta)}$$

$$\times H_{2l+p,2l+q}^{l+m,l+n}\left[z\middle| \begin{matrix}\Delta_l(-l, \alpha-l), (a_p, A_p), \Delta(l, \frac{1}{2}+\nu-\alpha)\\ \Delta(l, \beta), (b_q, B_q), \Delta_l(-l, \frac{1}{2}+\nu-\beta-l)\end{matrix}\right]$$

where $\mu \leqslant 0$, $R(\alpha + l\, b_j/B_j) > 0$ $(j = 1, \ldots, m)$; $R[\beta + l(1-a_j)/A_j] > 0$ $(j = 1, \ldots, n)$; $\lambda > 0$ and $|\arg z| < \pi\lambda/2$.

$$\int_0^1 t^{\alpha-1}(1-t)^{\beta-1}\, {}_4F_3\left[\begin{matrix}\alpha-\beta, 1+\dfrac{\alpha-\beta}{2}, \delta, -k; \dfrac{t}{t-1}\\ \dfrac{\alpha-\beta}{2}, 1+\alpha-\beta-\delta, 1+\alpha-\beta+k\end{matrix}\right]$$

$$\times H_{p,q}^{m,n}\left[\frac{z}{t^l(1-t)^l}\middle| \begin{matrix}(a_p, A_p)\\(b_q, B_q)\end{matrix}\right] dt$$

$$= \sqrt{\frac{\pi}{l}}\, 2^{1-\alpha-\beta} \frac{(1+\alpha-\beta)_k}{(1+\alpha-\beta-\delta)_k}$$

$$\times H_{3l+p,3l+q}^{2l+m,l+n}\left[2^{2l}z\middle| \begin{matrix}\Delta_l(-l, k-\beta-l+1), (a_p, A_p), \Delta(2l, \alpha+\beta)\\ \Delta(l, \alpha), \Delta_l(l, \beta-k), (b_q, B_q), \Delta_l(-l, 1-\beta-\delta-l)\end{matrix}\right]$$

where $R[\alpha - l(a_j - 1)/A_j] > 0$ $(j = 1, \ldots, n)$

$R[\beta - k - l(a_j - 1)/A_j) > 0$ $(j = 1, \ldots, n)$, $\mu \leqslant 0$, $\lambda > 0$

and $|\arg z| < \pi\lambda/2$.

(Banerji and Saxena, 1971)

2.13 Establish the following results.

$$\int_0^{\pi/2} (\sin\theta)^{\mu-\sigma-\frac{1}{2}} (\cos\theta)^{2\nu+1} {}_2F_1(-r, \mu+\nu+r+1; \mu+1; \sin^2\theta)$$

$$\times H^{m,n+1}_{p+2,q}\left[z \operatorname{cosec}^\delta\theta \,\middle|\, \begin{matrix} \left(\frac{1}{4}-\frac{\sigma}{2}-\frac{\mu}{2}, \frac{\delta}{2}\right), (a_p, A_p), \left(\frac{1}{4}-\frac{\sigma}{2}+\frac{\mu}{2}, \frac{\delta}{2}\right) \\ (b_q, B_q) \end{matrix}\right] d\theta$$

$$= \frac{\Gamma(\mu+1)\,\Gamma(\nu+r+1)}{2\Gamma(\mu+r+1)}$$

$$\times H^{m,n+1}_{p+1,q}\left[z \,\middle|\, \begin{matrix} \left(\frac{1}{4}-\frac{\sigma}{2}-\frac{\mu}{2}-r, \frac{\delta}{2}\right), (a_p, A_p), \left(\frac{5}{4}-\frac{\sigma}{2}+\frac{\mu}{2}+\nu+r, \frac{\delta}{2}\right) \\ (b_q, B_q) \end{matrix}\right]$$

where $\delta > 0$, $R(\mu) > -1$, $R(\nu) > -1$; $r = 0, 1, 2, \ldots$;

$R(\sigma + \delta \max (a_j - 1)/A_j) < 0$, $(j = 1, \ldots, n)$; $\lambda > 0$, $|\arg z| < \pi\lambda/2$

and $\sum_1^p A_j - \sum_1^q B_j \leqslant 0$;

$$\int_0^{\pi} (\sin\theta)^{\sigma}\, T_\nu^{-\mu}(\cos\theta)$$

$$\times H^{m,n+2}_{p+2,q}\left[z \sin^\delta\theta \,\middle|\, \begin{matrix} \left(-\frac{\sigma}{2}-\frac{\nu}{2}, \frac{\delta}{2}\right), \left(\frac{1-\sigma+\nu}{2}, \frac{\delta}{2}\right), (a_p, A_p) \\ (b_q, B_q) \end{matrix}\right] d\theta$$

$$= \frac{\pi}{2^\mu\, \Gamma\left(\frac{1+\mu-\nu}{2}\right) \Gamma\left(1+\frac{\mu}{2}+\frac{\nu}{2}\right)}$$

$$\times H^{m,n+2}_{p+2,q}\left[z \,\middle|\, \begin{matrix} \left(\frac{1-\sigma-\mu}{2}, \frac{\delta}{2}\right), \left(\frac{1-\sigma+\mu}{2}, \frac{\delta}{2}\right), (a_p, A_p) \\ (b_q, B_q) \end{matrix}\right]$$

where $\delta > 0$, $\lambda > 0$, $|\arg z| < \frac{\pi\lambda}{2}$ and $R\left(\sigma \pm \mu + \delta \min \frac{b_j}{B_j} + 1\right) > 0$, $(j = 1, \ldots, m)$, $\mu \leqslant 0$.

(Rathie, 1967)

REMARK. For integrals associated with Appell's function F_4 and $T_\nu^{-\mu}(x)$, see the work of Rathie (1967).

2.14 Establish the following formulas

$$\text{(i)} \int_0^\infty e^{2\sigma\theta} (\sinh\theta)^{2\rho-1} {}_2F_1(1/2+\rho-\sigma, \beta; \delta; 2e^{-\theta}\sinh\theta)$$

$$\times H_{p,q}^{m,n}\left[e^{2r\theta}(\sinh\theta)^{2r} z \,\middle|\, \begin{matrix}(a_p, A_p)\\(b_q, B_q)\end{matrix}\right] d\theta$$

$$= \frac{2^{-\rho-\sigma-\beta-(3/2)}\,\Gamma(\delta)}{\pi\,\Gamma(\delta-\beta)\,\Gamma(\frac{1}{2}+\rho-\sigma)}\, H_{p+4,q+4}^{m+4,n+2}\left[\frac{z}{2^{2r}}\,\middle|\,\chi\right],$$

where χ represents the following parameters

$$\left[\begin{matrix}(-\rho+1, r), \left(-\rho+\frac{1}{2}, r\right), (a_p, A_p), \left(-\rho+\frac{\delta}{2}, r\right), \left(-\rho+\frac{\delta}{2}+\frac{1}{2}, r\right)\\ \left(-\frac{\rho}{2}-\frac{\sigma}{2}+\frac{1}{4}, r\right), \left(-\frac{\rho}{2}-\frac{\sigma}{2}+\frac{3}{4}, r\right), \left(-\rho-\frac{\beta}{2}+\frac{\delta}{2}, r\right),\\ \left(-\rho-\frac{\beta}{2}+\frac{\delta+1}{2}, r\right), (b_q, B_q)\end{matrix}\right]$$

$$\lambda > 0,\ \mu \leqslant 0,\ |\arg z| < \lambda\pi/2,\ R(\delta-\beta-\rho+\sigma \quad 1/2) > 0,$$

$$R(\rho + r\, b_j/B_j) > 0\ (j = 1, \ldots, m);$$

$$R[\rho+\sigma-1/2+2r(a_j-1)/A_j] < 0\ (j = 1, \ldots, n).$$

$$\text{(ii)} \int_0^{\pi/2} e^{i(\alpha+\beta)\theta} (\sin\theta)^{\alpha-1} (\cos\theta)^{\beta-1}\, {}_2F_1(a, b; \beta; e^{i\theta}\cos\theta)\, d\theta$$

$$\times H_{p,q}^{m,n}\left[e^{ir(\theta-\pi/2)}(\sin\theta)^r z \,\middle|\, \begin{matrix}(a_p, A_p)\\(b_q, B_q)\end{matrix}\right] d\theta$$

$$= e^{i(\pi/2)\alpha}\,\Gamma(\beta)$$

$$H_{p+2,q+2}^{m,n+2}\left[z\,\middle|\,\begin{matrix}(1-\alpha, r), (1-\alpha-\beta+a+b, r), (a_p, A_p)\\ (b_q, B_q), (1-\alpha-\beta+a, r), (1-\alpha-\beta+b, r)\end{matrix}\right]$$

where $\mu\ 0 \leqslant,\ \lambda > 0,\ |\arg z| < \lambda\pi/2,\ R(\beta) > 0,\ R(\beta-a-b) > 0$

and $R(\alpha + r\, b_j/B_j) > 0 \quad (j = 1, \ldots, m).$

REMARK. For the proofs of the integrals of this section and for further integrals of this type, see the work of Anandani (1971b).

.15 Prove that the Mellin-Barnes integral representation for Srivastava and Daoust's function can be given in the following manner.

$$S_{C:D;D'}^{A:B;B'}\left(\begin{matrix}x\\y\end{matrix}\,\middle|\,\begin{matrix}[(a):\theta, \theta']:[(b):\beta], [(b');\beta']\\ [(c):\phi, \phi']:[(d):\delta]; [(d'):\delta']\end{matrix}\right)$$

$$= -\frac{1}{4\pi^2}\int_{-i\infty}^{-i\infty}\int_{-i\infty}^{i\infty} \chi_1^*(s_1, s_2)\, \chi_1^*(s_1)\, \chi_2^*(s_2)\,(-x)^{s_1}(-y)^{s_2}\, ds_1\, ds_2$$

where
$$\chi^*(s_1, s_2) = \frac{\prod_{j=1}^{A} \Gamma(a_j + s_1\theta_j + s_2\theta'_j)}{\prod_{j=1}^{C} \Gamma(c_j + s_1\phi_j + s_2\phi'_j)},$$

$$\chi^*(s_1) = \frac{\prod_{j=1}^{B} \Gamma(b_j + s_1\beta_j)\,\Gamma(-s_1)}{\prod_{j=1}^{D} \Gamma(d_j + s_1\delta_j)},$$

$$\chi^*(s_2) = \frac{\prod_{j=1}^{B'} \Gamma(b'_j + s_2\beta'_j)\,\Gamma(-s_2)}{\prod_{j=1}^{D'} \Gamma(d'_j + s_2\delta'_j)},$$

where the paths of integration are indented, if necessary, to ensure that the poles of gamma functions with negative s_1 and s_2 are separated from the poles of all the gamma functions with positive s_1 and s_2 and the poles of the integrand are assumed to be simple. (Saxena and Verma, 1977)

REMARK. Similar Mellin-Barnes integral representation for the generalized Lauricella functions due to Srivastava and Daoust can be found in the work of Saxena (1977).

2.16 Prove that

$$\int_0^\infty\int_0^\infty x^{\alpha-1} y^{\beta-1}(Ax^\gamma + By^\delta)^\sigma H_{p,q}^{m,n}\left[Wx^s y^k (Ax^\gamma + By^\delta)^r \,\middle|\, \begin{matrix}(a_p, A_p)\\(b_q, B_q)\end{matrix}\right]$$
$$\times H_{P,Q}^{M,N}\left[\lambda(Ax^\gamma + By^\delta)\,\middle|\, \begin{matrix}(c_P, C_P)\\(d_Q, D_Q)\end{matrix}\right] dx\, dy$$
$$= \frac{A^{-\alpha/\gamma} B^{-\beta,\delta} \lambda^{-\eta}}{\gamma^\delta} H_{P+Q+2,Q+P+1}^{M+N,N+M+2}\left[\frac{WA^{-s/\gamma} B^{-k/\delta}}{\lambda^\theta}\,\middle|\, \chi^*\right]$$

where

$$\chi^* = \left[\begin{matrix}\left(1-\frac{\alpha}{\gamma}, \frac{s}{\gamma}\right), \left(1-\frac{\beta}{\delta}, \frac{k}{\delta}\right), (\nu_Q;\, \theta D_Q), (a_p, A_p)\\ (\sigma_P, \theta C_P), (b_q, B_q), (1-\eta+\sigma, \theta - r)\end{matrix}\right]$$

where m, n, p, q and M, N, P, Q are integers, such that $0 \leqslant m \leqslant q$, $0 \leqslant n \leqslant p$, $0 \leqslant M \leqslant Q$, $0 \leqslant N \leqslant P$; A, B; γ, δ, s, k, r and $A_1, \ldots, A_p$,

$B_1, \ldots, B_q, C_1, \ldots, C_P, D_1, \ldots, D_Q$ are all positive, $R(\alpha) > 0$, $R(\beta) > 0$ and

$$-\min_{1 \leqslant j \leqslant M} R\left(\frac{d_j}{D_j}\right) - \min_{1 \leqslant j \leqslant m} \left\{R\left(\frac{\theta b_j}{B_j}\right)\right\}$$

$$< R(n) < \min R\left\{\left(\frac{1-c_j}{C_j}\right)\right\} + \min\left\{\theta R\left(\frac{1-a_h}{A_h}\right), \frac{\theta}{s} R(\alpha), \frac{\theta}{K} R(\beta)\right\}$$

for $h = 1, \ldots, n$; $\theta = r + \frac{s}{\gamma} + \frac{k}{\delta}$, $\eta = \sigma + \frac{\alpha}{\gamma} + \frac{\beta}{\delta}$, $\nu_j = 1 - d_j - \eta D_j$, $j = 1, \ldots, Q$, $\sigma_j = 1 - c_j - \eta C_j$, $j = 1, \ldots, P$; $|\arg W| < 1/2\ \pi\psi_1$ and $|\arg \lambda| < \pi\psi_2/2$, where

$$\psi_1 = \sum_1^n A_j - \sum_{n+1}^p A_j + \sum_1^m B_j - \sum_{m+1}^q B_j > 0;$$

$$\psi_2 = \sum_1^N C_j - \sum_{N+1}^P C_j + \sum_1^M D_j - \sum_{M+1}^Q D_j > 0.$$

(Srivastava, Gupta and Handa, 1975)

REMARK. Recently Joshi, C.M. and Prajapat. M.L. (1977) have derived the convergence conditions of H-function of two variables on employing the method used by MacRobert, T.M. (1962) for his E-function. It is also learnt that Buschman, R.G. has given a detailed analysis of convergence conditions of H-function of two variables in a series of papers, one of which is "H-function of two variables–I", to appear in Indian Journal of Mathematics, Vol. 20, (P.L. Bhatnagar Memorial Volume).

CHAPTER 3

Finite and Infinite Series for the *H*-function

This chapter deals with the finite and infinite summation formulae for the *H*-functions. The results of this chapter may find applications in various problems arising in physical, biological and statistical sciences. Certain known as well as some new results on the summation formulae for the Bessel functions, Whittaker functions and hypergeometric func tions can be derived from the main results of this chapter by specializing the parameters suitably.

The results of this chapter can be obtained by employing the definition (1.1.1) and adopting the methods of summation presented in the next section. A detailed discussion of all these methods can be found in a recent monograph by Mathai and Saxena (1973a). Hence the details of the proofs are omitted.

Finite and infinite series expansions for the *H*-function are derived by Anandani (1967, 1968a, 1969b, 1969c, 1969f, 1969g, 1970a, 1970f, 1970g, 1970h, 1970i, 1970k, 1971a, 1971c, 1971d, 1972a and 1972b), Bajpai (1969a, 1969b, 1969c, 1970, 1970a and 1970b), Olkha (1970), Parashar (1967), Shah (1969, 1969a, 1969b, 1969d), Goyal, A.N. (1969), Goyal, A.N and Goyal, G. K. (1967a), Saxena, (1971a), Saxena and Mathur (1971), Srivastava, H.M. and Daoust, M.C. (1969a), Singh, F. (1972), Singh, F. and Varma, R.C. (1972), Lawrynowicz (1969), Skibinski (1970), Jain, R.N. (1969), Taxak (1971b, 1973) and others. Since most of these expansions are developed on parallel lines to that of the series expansions for a Meijer's *G*-function discussed by Mathai and Saxena (1973a), only those cases which are not covered there will be treated here.

It is interesting to remark here that the existing results on finite and infinite series of *G*-function of one variable are extended to *G*-function of two variables, whose definitions are given in Section 2.1.1, by many workers notably by Sharma, B.L., Agarwal, R.P., Bora and Kalla, Kalla, Saxena, R.K., Verma, R.U., Shah, M., Anandani, Munot and Kalla and other workers, which can be seen from the papers cited in the bibliography.

In the present chapter we will deal with only the series expansions of an H-function. Some series expansions of H-function of two variables are, however, given in the exercises.

A general expansion of an H-function, when the poles of the integrand of the H-function differ in any manner, due to Mathai (1973b), will be given in Section 3.7.

3.1 METHODS OF SUMMATION

The existing methods, enumerated below, for the summation of generalized hypergeometric functions, involve the substitution of the Mellin-Barnes contour (1.1.1) for the given generalized hypergeometric function, interchange of the order of summation and integration and following one of these methods:

METHOD 1. In certain cases, by the use of the multiplication formula for the gamma functions and its various properties, the given series can be summed.

METHOD 2. In case the generalized hypergeometric function involves a given orthogonal polynomial or a Bessel function $J_\nu(x)$, then the given series can be summed by using the orthogonality properties of that polynomial or $J_\nu(x)$. The orthogonality properties of the various polynomials and the Bessel function $J_\nu(x)$ can be found in Section 3.1.1 and 3.1.2 of the monograph by Mathai and Saxena (1973a).

METHOD 3. On employing the various known summation theorems for the hypergeometric functions with specialized arguments such as the summation theorems for ${}_2F_1(+1)$, ${}_2F_1(-1)$, ${}_2F_1(\frac{1}{2})$, ${}_3F_2(+1)$, ${}_4F_3(+1)$, etc. in writing the value of the inner integrals involved in the process, a number of interesting and useful results can be derived.

METHOD 4. By making use of the technique of Laplace transform and its inverse to effect the induction with respect to certain parameters in the summation formulae under consideration, a number of summation formulae can be established without difficulty.

METHOD 5. If we know a finite or an infinite summation formula for any given elementary special function in terms of products of gamma functions or Neumann type of expansions for the given special function, then by employing suitable integrals of products of the given function and the given generalized hypergeometric function we can prove a number of finite or infinite summation formulae.

METHOD 6. On proving the equality of the coefficients of the general term on either side, a large number of summation formulae can be established.

3.2 FOURIER SERIES FOR THE *H*-FUNCTION

The following two results (3.2.1) and (3.2.2) given below follow as special cases of an integral evaluated by Rathie, P.N. (1967), which will be needed in proving the results that follow.

$$\int_0^{\pi} \sin(2u+1)\theta\,(\sin\theta)^{1-2\zeta}\, H_{p,q}^{m,n}\left[z(\sin\theta)^{-2h}\,\middle|\,\begin{matrix}(a_p, A_p)\\(b_q, B_q)\end{matrix}\right] d\theta \tag{3.2.1}$$

$$= \sqrt{\pi}\, H_{p+2,q+2}^{m+1,n+1}\left[z\,\middle|\,\begin{matrix}(1-\zeta-u, h), (a_p, A_p), (2-\zeta+u, h)\\(\tfrac{3}{2}-\zeta, h), (b_q, B_q), (1-\zeta, h)\end{matrix}\right]$$

where $R[3-2\zeta+2h(1-a_j)/A_j]>0$ $(j=1,\ldots,n)$; $h>0$, $\lambda>0$, $\mu\leqslant 0$, $|\arg z|<\lambda\pi/2$, $u=0,1,\ldots$; λ and μ are defined in (2.6.5).

$$\int_0^{\pi} \cos u\theta \left(\sin\frac{\theta}{2}\right)^{1-2\zeta} H_{p,q}^{m,n}\left[z\left(\sin\frac{\theta}{2}\right)^{-2h}\,\middle|\,\begin{matrix}(a_p, A_p)\\(b_q, B_q)\end{matrix}\right] d\theta \tag{3.2.2}$$

$$= \sqrt{\pi}\, H_{p+2,q+2}^{m+1,n+1}\left[z\,\middle|\,\begin{matrix}(1-\zeta-u, h), (a_p, A_p), (1-\zeta+u, h)\\(\tfrac{1}{2}-\zeta, h), (b_q, B_q), (1-\zeta, h)\end{matrix}\right],$$

where $R[1-2\zeta+2h(1-a_j)/A_j]>0$ $(j=1,\ldots,n)$; $h>0$, $\lambda>0$,

$$|\arg z|<\pi\lambda/2;\ u=0,1,2,\ldots;\ \text{and}\ \mu\leqslant 0.$$

Bajpai (1969b) has given the following Fourier series for the *H*-function, which readily follows from the integrals (3.2.1) and (3.2.2) given above, on using the orthogonality property of sine functions. For details, see Bajpai (1969b, pp 705-6).

$$(\sin\theta)^{1-2\zeta}\, H_{p,q}^{m,n}\left[z(\sin\theta)^{-2h}\,\middle|\,\begin{matrix}(a_p, A_p)\\(b_q, B_q)\end{matrix}\right] \tag{3.2.3}$$

$$= \frac{2}{\sqrt{\pi}} \sum_{r=0}^{\infty} \sin(2r+1)\theta$$

$$\times\, H_{p+2,q+2}^{m+1,n+1}\left[z\,\middle|\,\begin{matrix}(1-\zeta-r, h), (a_p, A_p), (2-\zeta+r, h)\\(\tfrac{3}{2}-\zeta, h), (b_q, B_q), (1-\zeta, h)\end{matrix}\right],$$

where $R(1-2\zeta)\geqslant 0$, $0\leqslant\theta\leqslant\pi$; $h>0$, $\mu\leqslant 0$, $\lambda>0$ and $|\arg z|<\pi\lambda/2$.

$$\left(\sin\frac{\theta}{2}\right)^{-2\zeta} H_{p,q}^{m,n}\left[z\left(\sin\frac{\theta}{2}\right)^{-2h}\,\middle|\,\begin{matrix}(a_p, A_p)\\(b_q, B_q)\end{matrix}\right] \tag{3.2.4}$$

$$= \frac{1}{\sqrt{\pi}}\, H_{p+1,q+1}^{m+1,n}\left[z\,\middle|\,\begin{matrix}(a_p, A_p), (1-\zeta, h)\\(\tfrac{1}{2}-\zeta, h), (b_q, B_q)\end{matrix}\right]$$

$$+\frac{2}{\sqrt{\pi}} \sum_{r=1}^{\infty} \cos(r\theta)\, H_{p+2,q+2}^{m+1,n+1}\left[z\,\middle|\,\begin{matrix}(1-\zeta-r, h), (a_p, A_p), (1-\zeta+r, h)\\(\tfrac{1}{2}-\zeta, h), (b_q, B_q), (1-\zeta, h)\end{matrix}\right],$$

where $0 \leqslant \theta \leqslant \pi$, $R(\zeta) \leqslant 0$, $h > 0$, $\mu \leqslant 0$, $\lambda > 0$ and $|\arg z| < \pi\lambda/2$.

REMARK. When $\zeta = 0$, (3.2.3) and (3.2.4) reduce to the results given by Anandani (1968b), which itself are the generalizations of the results due to Parashar (1967).

Reducing the H-function to a G-function in (3.2.3), we find that

$$(\sin\theta)^{1-2\zeta}\, G^{m,n}_{p,q}\left[z\,(\sin\theta)^{-2h}\,\middle|\,\begin{matrix} a_p \\ b_q \end{matrix}\right] \tag{3.2.5}$$

$$= \frac{2}{\sqrt{\pi h}} \sum_{r=0}^{\infty} \sin(2r+1)\theta$$

$$\times G^{m+h\; n+h}_{p+2h,q+2h}\left[z\,\middle|\,\begin{matrix} \Delta(h, 1-\zeta-r),\, a_p,\, \Delta(h, 2-\zeta+r) \\ \Delta(h, \frac{3}{2}-\zeta),\, b_q,\, \Delta(h, 1-\zeta) \end{matrix}\right]$$

where h is a positive integer, $m+n > \frac{p}{2}+\frac{q}{2}$, $R(\zeta) \leqslant \frac{1}{2}$, $0 \leqslant \theta \leqslant \pi$, $|\arg z| < \left(m+n-\frac{p}{2}-\frac{q}{2}\right)\pi$.

For $h = 1$, (3.2.5) gives rise to a result due to Jain, R.N. (1965, p. 103, 2.4).

Parashar (1967) has established the following expansion formula.

$$\sqrt{\pi}\, H^{m,n}_{p,q}\left[x\cos^2\left(\frac{\theta}{2}\right)\middle|\,\begin{matrix} (a_p, A_p) \\ (b_q, B_q) \end{matrix}\right] \tag{3.2.6}$$

$$= H^{m,n+1}_{p+1,q+1}\left[x\,\middle|\,\begin{matrix} (\frac{1}{2}, 1), (a_p, A_p) \\ (b_q, B_q), (0, 1) \end{matrix}\right]$$

$$+ 2\sum_{r=1}^{\infty} \cos(r\theta)\, H^{m,n+2}_{p+2,q+2}\left[x\,\middle|\,\begin{matrix} (\frac{1}{2}, 1), (0, 1), (a_p, A_p) \\ (b_q, B_q), (-r, 1), (r, 1) \end{matrix}\right]$$

To establish the result (3.2.6) we require the following results.

$$\int_0^{\pi/2} \cos^{\alpha}\theta \cos\beta\theta\, d\theta = \frac{\pi\,\Gamma(\alpha+1)}{2^{\alpha+1}\,\Gamma\left(\frac{\alpha+\beta}{2}+1\right)\Gamma\left(\frac{\alpha-\beta}{2}+1\right)} \tag{3.2.7}$$

where $R(\alpha) > -1$.

$$\sqrt{\pi}\,\frac{\Gamma(s+1)}{\Gamma(s+\frac{1}{2})}\left(\cos\frac{\theta}{2}\right)^{2s} = 1 + 2\sum_{r=1}^{\infty} \frac{(-1)^r(-s)_r}{(s+1)_r}\cos(r\theta) \tag{3.2.8}$$

where $R(s) > -1/2$ and $0 \leqslant \theta \leqslant \pi$.

Equation (3.2.8) can be proved with the help of (3.2.7). But (3.2.6) can be established by expressing the H-function as a Mellin-Barnes integral, substituting the series for $\cos^{2s}(\theta/2)$ from (3.2.8) and changing the order of integration and summation, which is permissible under the conditions stated.

The results given by Saxena (1971, pp. 6–7) are as follows. These results can be proved by making use of the integrals (2.8.5), (2.8.6) and (2.8.7) and the orthogonality property of the sine functions.

$$(\sin \delta\theta)^{\gamma-2} \exp(-\mu\theta)\, H_{p,q}^{m,n}\left[z (\sin \delta\theta)^{2h} \middle| \begin{matrix} (a_p, A_p) \\ (b_q, B_q) \end{matrix} \right] \tag{3.2.9}$$

$$= \frac{1}{2^{\gamma-2}} \sum_{u=1}^{\infty} \exp\left(-\frac{\pi\mu}{2u}\right) \sin(u\theta)\, H_{p+1,q+2}^{m,n+1}\left[\frac{z}{4^h} \middle| \chi \right]$$

where χ stands for the parameters

$$\left[\begin{matrix} (1-\gamma, 2h), (a_p, A_p) \\ (b_q, B_q), \left(\dfrac{u \pm i\mu - \gamma u}{2u}, h \right) \end{matrix} \right]$$

where $h > 0$, $R(\gamma + 2h\, b_j/B_j) > 0$, $\sum_1^q B_j - \sum_1^p A_j \geqslant 0$, $\lambda > 0$,

$|\arg z| < \pi\lambda/2$ and $0 < \theta < \pi/\delta$.

$$(\sin \delta\theta)^{\gamma-2} e^{-\mu\theta}\, H_{p,q}^{m,n}\left[z e^{i2h\delta\theta} \middle| \begin{matrix} (a_p, A_p) \\ (b_q, B_q) \end{matrix} \right] \tag{3.2.10}$$

$$= \frac{1}{2^{\gamma-2}} \sum_{u=1}^{\infty} \exp\left(-\frac{\pi\mu}{2u}\right) \sin(u\theta)$$

$$\times H_{p+1,q+1}^{m,n}\left[z e^{i\pi h} \middle| \begin{matrix} (a_p, A_p), \left(\left(\dfrac{u - i\mu + \gamma u}{2u}, h \right) \right) \\ (b_q, B_q), \left(\left(\dfrac{u - i\mu - \gamma u}{2u}, h \right) \right) \end{matrix} \right]$$

where $h > 0$, $R(\gamma) > 0$, $\sum_1^q B_j - \sum_1^p A_j \geqslant 0$; $\lambda > 0$

$|\arg z| < \pi\lambda/2$ and $0 < \theta < \pi/\delta$.

$$(\sin \delta\theta)^{\gamma-2} e^{-\mu\theta}\, H_{p,q}^{m,}\left[z (\sin \delta\theta)^{2l} e^{2ih\delta\theta} \middle| \begin{matrix} (a_p, A_p) \\ (b_q, B_q) \end{matrix} \right] \tag{3.2.11}$$

$$= \frac{1}{2^{\gamma-2}} \sum_{u=0}^{\infty} \sin(u\theta) \exp\left(-\frac{\pi\mu}{2u}\right)$$

$$\times H_{p+1,q+2}^{m,n+1}\left[\frac{z e^{i\pi h}}{4^l} \middle| \chi_1 \right]$$

where χ_1 denotes the parameters

$$\left[\begin{matrix}(1-\gamma, 2l), (a_p, A_p)\\ (b_q, B_q),\ (u-i\mu-\gamma u)/2u, l+h),\ (u+i\mu-\gamma u)/2u, l-h)\end{matrix}\right]$$

provided that $l-h>0$, $R(\gamma+2l\, b_j/B_j)>0$ $(j=1,\ldots,m)$;

$$\sum_1^q B_j - \sum_1^p A_j \geqslant 0,\ \lambda>0,\ |\arg z| < \pi\lambda/2 \text{ and } 0<\theta<\pi/\delta.$$

When $h=1$, $A_j=B_h=1$ $(j=1,\ldots,p;\ h=1,\ldots,q)$, then (3.2.9) and (3.2.10) yield the following series expansions for the Meijer's G-function.

$$(\sin\delta\theta)^{\gamma-2}\, e^{-\mu\theta}\, G_{p,q}^{m,n}\left(z\sin^2\delta\theta\,\middle|\begin{matrix}a_p\\ b_q\end{matrix}\right) \tag{3.2.12}$$

$$=\frac{1}{2\sqrt{\pi}}\sum_{u=1}^{\infty}\exp\left(-\frac{\pi\mu}{2u}\right) G_{p+2,q+2}^{m,n+1}\left[\frac{z}{4}\,\middle|\begin{matrix}\frac{1-\gamma}{2}, \frac{2-\gamma}{2}, a_p\\ b_q, \frac{u\pm i\mu+\gamma u}{2u}\end{matrix}\right]$$

where $R(\gamma+2\,b_j)>0$, $(j=1,\ldots,m)$; $m+n>\frac{1}{2}p+\frac{1}{2}q$,
$|\arg z| < \left(m+n-\frac{p}{2}-\frac{q}{2}\right)\pi$ and $0<\theta<\frac{\pi}{\delta}$.

$$(\sin\delta\,\theta)^{\gamma-2}\, e^{-\mu\theta}\, G_{p,q}^{m,n}\left[ze^{i2\delta\theta}\,\middle|\begin{matrix}a_p\\ b_q\end{matrix}\right] \tag{3.2.13}$$

$$=\frac{1}{2^{\gamma-2}}\sum_{u=1}^{\infty}\sin(u\,\theta)\exp\left[-\frac{\pi\,\mu}{2\,u}\right]$$

$$\times\, G_{p+1,q+1}^{m,n}\left[e^{i\pi}z\,\middle|\begin{matrix}a_p, \frac{u-i\mu+\gamma\mu}{2u}\\ b_q, \frac{u-i\mu-\gamma\mu}{2u}\end{matrix}\right]$$

where

$$R(\gamma)>0,\ |\arg z| < \left(m+n-\frac{p}{2}-\frac{q}{2}-1\right)\pi,$$

$$m+n>\frac{p}{2}+\frac{q}{2}+1,\ 0<\theta<\frac{\pi}{\delta}.$$

As a consequence of the results (3.2.3) and (3.2.4) with $h=1$ and $\zeta=0$ and (3.2.6), we have the following integrals.

$$\int_0^{\pi} H_{p,q}^{m,n}\left[x\cos^2\frac{\theta}{2}\,\middle|\begin{matrix}(a_p, A_p)\\ (b_q, B_q)\end{matrix}\right]\cos(r\theta)\, d\theta \tag{3.2.14}$$

$$= \sqrt{\pi}\; H^{m,n+2}_{p+2,q+2}\left[x\,\middle|\,\begin{matrix}(\tfrac{1}{2},1),(0,1),(a_p,A_p)\\(b_q,B_q),(-r,1),(r,1)\end{matrix}\right].$$

$$\int_0^{\pi} \sin((2r+1)\theta)\sin\theta\; H^{m,n}_{p,q}\left[\frac{x}{\sin^2\theta}\,\middle|\,\begin{matrix}(a_p,A_p)\\(b_q,B_q)\end{matrix}\right]d\theta \tag{3.2.15}$$

$$= \sqrt{\pi}\; H^{m+1,n+1}_{p+2,q+2}\left[x\,\middle|\,\begin{matrix}(1-r,1),(a_p,A_p),(2+r,1)\\(3/2,1),(b_q,B_q),(1,1)\end{matrix}\right].$$

$$\int_0^{\pi} \cos(r\theta)\; H^{m,n}_{p,q}\left[\frac{x}{\sin^2(\theta/2)}\,\middle|\,\begin{matrix}(a_p,A_p)\\(b_q,B_q)\end{matrix}\right]d\theta \tag{3.2.16}$$

$$= \sqrt{\pi}\; H^{m+1,n+1}_{p+2,q+2}\left[x\,\middle|\,\begin{matrix}(1-r,1),(a_p,A_p),(1+r,1)\\(\tfrac{1}{2},1),(b_q,B_q),(1,1)\end{matrix}\right];$$

which can also be written as follows:

$$\int_0^{\pi} \cos(2r\theta)\; H^{m,n}_{p,q}\left[\frac{x}{\sin^2\theta}\,\middle|\,\begin{matrix}(a_p,A_p)\\(b_q,B_q)\end{matrix}\right]d\theta \tag{3.2.17}$$

$$= \sqrt{\pi}\; H^{m+1,n+1}_{p+2,q+2}\left[x\,\middle|\,\begin{matrix}(1-r,1),(a_p,A_p),(1+r,1)\\(\tfrac{1}{2},1),(b_q,B_q),(1,1)\end{matrix}\right].$$

On using the formula

$$\tfrac{1}{2}[\cos(2r\theta)-\cos((2r+2)\theta)] = \sin((2r+1)\theta)\sin\theta,$$

(3.2.17) leads to the recurrence relation,

$$2\; H^{m+1,n+1}_{p+2,q+2}\left[x\,\middle|\,\begin{matrix}(1-r,1),(a_p,A_p),(2+r,1)\\(3/2,1),(b_q,B_q),(1,1)\end{matrix}\right]$$

$$= H^{m+1,n+1}_{p+2,q+2}\left[x\,\middle|\,\begin{matrix}(1-r,1),(a_p,A_p),(2+r,1)\\(\tfrac{1}{2},1),(b_q,B_q),(1,1)\end{matrix}\right]$$

On changing H-functions into MacRobert's E-functions from the relation (1.7.10) it gives rise to the results given earlier by MacRobert (1961).

Remark. For further Fourier series for the H-function, see the work of Anandani (1968b) and Shah (1969). Fourier series for H-function of two variables is given by Gupta, S.D. (1973a).

3.3 EXPANSIONS OF CERTAIN SPECIAL FUNCTIONS IN SERIES OF LAGUERRE POLYNOMIALS

On using the integral (2.10.1) and the orthogonality property of the Laguerre polynomials, Shah (1969a) has derived an expansion formula for an H-function in terms of an infinite series of products of a related H-function and Laguerre polynomials in the following form:

$$x^{\gamma} H_{p,q}^{m,n}\left[z\lambda^{\delta}\left|\begin{matrix}(a_p, A_p)\\(b_q, B_q)\end{matrix}\right.\right] = (2\pi)^{(1-\delta)/2}\, \delta^{\gamma+\sigma+1/2} \tag{3.3.1}$$

$$\times \left[\sum_{r=0}^{\infty} \frac{(-1)^r \delta^r}{\Gamma(\sigma+r+1)} L_r^{(\sigma)}(x)\right.$$

$$\left.\times H_{p+2\delta,q+\delta}^{m,n+2\delta}\left[z\delta^{\delta}\left|\begin{matrix}(\Delta(\delta,-\gamma-\sigma),1),(\Delta(\delta,-\gamma),1),(a_p,A_p)\\(b_q,B_q),(\Delta(\delta,-\gamma+r),1)\end{matrix}\right.\right]\right]$$

where δ is a positive integer

$$\sum_1^p A_j - \sum_1^q B_j \leqslant 0,\ |\arg z| < \lambda\pi/2$$

$$\lambda = \sum_1^n A_j - \sum_{n+1}^p A_j + \sum_1^m B_j - \sum_{m+1}^q B_j > 0,$$

$$R(\gamma + \delta\, b_j/B_j) > -1\ (j = 1,\ldots, m).$$

Below we list some interesting particular cases of (3.3.1) which may be found useful in various problems in pure and applied mathematics.

$$x^a J_{\nu}^{\mu}(zx) = \sum_{r=0}^{\infty} \frac{(-1)^r}{\Gamma(\sigma+r+1)}$$

$$H_{2,3}^{1,2}\left[z\left|\begin{matrix}(-a-\sigma,1),(-a,1)\\(0,1),(-\nu,\mu),(-a+r,1)\end{matrix}\right.\right] L_r^{(\sigma)}(x) \tag{3.3.2}$$

where $|\arg z| < \pi/2$ and $R(\gamma) > -1$.

$$2x^{a+b/2} z^{b/2} K_{\nu}(2\sqrt{zx}) \tag{3.3.3}$$

$$= \sum_{r=0}^{\infty} \frac{(-1)^r}{\Gamma(\sigma+r+1)} G_{2,k}^{2,1}\left(z\left|\begin{matrix}-a-\sigma,-a\\ \frac{b+\nu}{2},\frac{b-\nu}{2},r-a\end{matrix}\right.\right) L_r^{(\sigma)}(x)$$

where $|\arg z| < \pi$ and $R\left(a+\frac{b}{2}\pm\frac{\nu}{2}\right) > -1$.

$$2\pi^{1/2} x^a I_b(\sqrt{zx}) K_b(\sqrt{(zx)}) \tag{3.3.4}$$

$$= \sum_{r=0}^{\infty} \frac{(-1)^r}{\Gamma(\sigma+r+1)} G_{3,4}^{2,3}\left(z\left|\begin{matrix}-a-\sigma,-a,\frac{1}{2}\\ b,0,-b,-a+r\end{matrix}\right.\right) L_r^{(\sigma)}(x),$$

where $|\arg z| < \pi$, $R(a \pm b) > -1$,

$$x^{a+1} z^b e^{-xz/2} W_{\lambda,\mu}(zx) \tag{3.3.5}$$

$$= \sum_{r=0}^{\infty} \frac{(-1)^r}{\Gamma(\sigma+r+1)} L_r^{(\sigma)}(x)$$

$$\times G_{2,3}^{2,2}\left(z \,\middle|\, \begin{matrix} -a-\sigma, -a, b-\lambda+1 \\ \frac{1}{2}+b+\mu, \frac{1}{2}+b-\mu, -a+r \end{matrix}\right)$$

where $R(a \pm \mu + b) > -3/2$.

$$x^{a-1/2}\pi^{1/2}z^{-1/2}W_{c,d}(2\sqrt{(zx)})\,W_{-c,d}[2\sqrt{(zx)}] \tag{3.3.6}$$

$$= \sum_{r=0}^{\infty} \frac{(-1)^r}{\Gamma(\sigma+r+1)} L_r^{(\sigma)}(x)\; G_{4,5}^{4,2}\left(z \,\middle|\, \begin{matrix} -a-\sigma, -a, \frac{1}{2}+c, \frac{1}{2}-c \\ 0, \frac{1}{2}, d, -d, -a+r \end{matrix}\right)$$

where $R(a \pm d) > -1$, $|\arg z| < \pi/2$.

3.4 SERIES AND INTEGRAL OF *H*-FUNCTION ASSOCIATED WITH GEGENBAUER POLYNOMIALS

Shah (1972f) has shown that

$$\sum_{u=0}^{\infty} H_{p+3,q+3}^{m+2,n+1}\left[4z \,\middle|\, \begin{matrix} (2-\xi-u, 1), (a_p, A_p), (1, 1), (l+u+2, 1) \\ (l+2, 2), (b_q, B_q), (2-\xi, 1) \end{matrix}\right]$$

$$\times \frac{2^{2\xi-2}\,\Gamma(\xi)\,(l+2u+\xi)\,(l+2u)!\,\Gamma(l+u+\xi)}{l!\,u!\,\Gamma(l+2u+2\xi)} \tag{3.4.1}$$

$$\times C_{l+2u}^{\xi}(\cos\theta)\,(\sin\theta)^{2\xi}$$

$$= \sin^2\theta \sum_{k=0}^{l} \frac{(-l)_k}{k!\,l!} \cos((l-2k)\theta)$$

$$\times H_{p+3,q+3}^{m+2,n+1}\left[\frac{z}{\sin^2\theta} \,\middle|\, \begin{matrix} (l+1, 1), (a_p, A_p), (1, 1), (1, 1) \\ (1+k, 1), (1+l, 1), (b_q, B_q), (l-k+1, 1) \end{matrix}\right]$$

where $0 \leqslant \theta \leqslant \pi$, $R[l+2(1-a_j)/A_j] > 0$ $(j=1,\ldots,n;$ $R[b_h/B_h] > -1$ $(h=1,\ldots,m)$, $\sum_1^p A_j - \sum_1^q B_j \leqslant 0$;

$$\lambda = \sum_1^n A_j - \sum_{n+1}^p A_j + \sum_1^m B_j - \sum_{m+1}^q B_j > 0 \text{ and } |\arg z| < \pi\lambda/2.$$

He has evaluated the integral

$$\sum_{k=0}^{l} \frac{(-l)_k}{k!\,l!} \int_0^{\pi} H_{p+3,q+3}^{m+2,n+1}\left[\frac{z}{\sin^2\theta} \,\middle|\, \begin{matrix} (l+1, 1), (a_p, A_p), (1, 1), (1, 1) \\ (1+k, 1), (1+l, 1), (b_q, B_q), (l-k+1, 1) \end{matrix}\right]$$

$$\times \sin^2\theta \cos(l-2k)\theta\; C_{l+2\gamma}^{\xi}(\cos\theta)\,d\theta \tag{3.4.2}$$

$$= \frac{\pi}{2} \frac{\Gamma(l+\gamma+\xi)}{l!\,\gamma!\,\Gamma(\xi)}$$

$$\times\, H^{m+2,n+1}_{p+3,q+3}\left[4z\left|\begin{matrix}(2-\xi-\gamma,1),(a_p,A_p),(1,1),(l+\gamma+2,1)\\(l+2,2),(b_q,B_q),(2-\xi,1)\end{matrix}\right.\right]$$

where $0 \leqslant \theta \leqslant \pi$ and $\gamma = 0, 1, 2 \ldots .$

In proving the above series and the integral the following result due to Askey (1965) and the orthogonality properties of Gegenbauer polynomials are used,

$$(\sin\theta)^{2\gamma}\, C_l^{\gamma}(\cos\theta) \tag{3.4.3}$$

$$= \sum_{k=0}^{\infty} A_{k,l}^{\gamma,\xi}\; C^{\xi}_{l+2k}(\cos\theta)\;(\sin\theta)^{2\xi},$$

where

$$A_{k,l}^{\gamma,\xi} = \frac{2^{2\xi-2\gamma}\,\Gamma(\xi)\,(l+2k+\xi)\,(l+2k)!\,\Gamma(l+k+\xi)\,\Gamma(k+\xi-\gamma)}{l!\,k!\,\Gamma(\gamma)\,\Gamma(\xi-\gamma)\,\Gamma(l+k+\gamma+1)\,\Gamma(l+2k+2\xi)};$$

$$A_{k,l}^{\gamma,\xi} > 0 \text{ and } \frac{\xi-1}{2} < \gamma < \xi.$$

$$\int_{-1}^{1} (1-x^2)^{\gamma-1/2}\, C_l^{\gamma}(x)\, C_m^{\gamma}(x)\, dx \tag{3.4.4}$$

$$= \begin{cases} 0 \text{ if } m \neq l, \\ \dfrac{(2\gamma)_l\,\Gamma(\frac{1}{2})\,\Gamma(\gamma+\frac{1}{2})}{l!\,(\gamma+l)\,\Gamma(\gamma)}, \text{ if } m = l \end{cases}$$

where $R(\gamma) > -\frac{1}{2}$ and $C_l^{\gamma}(x)$ denotes a Gegenbauer (ultraspherical) polynomial.

Remark. When $\xi = 1$, (3.4.3) reduces to a known series given by Szegö (1939).

3.5 SERIES FOR THE *H*-FUNCTION INVOLVING GENERALIZED LEGENDRE'S ASSOCIATED FUNCTIONS

The following expansion formula is due to Anandani (1970g) and it can be established by using the orthogonality property of generalized Legendre's associated function given by Meulenbeld and Robin (1961, p. 340, 26 and 27) and an integral evaluated by Anandani (1969e).

We have

$$(1-x)^{\rho-m/2}(1+x)^{\sigma+n/2}\, H^{p,q}_{r,s}\left[z(1-x)^{\nu}(1+x)^{\delta}\left|\begin{matrix}(a_r,A_r)\\(b_s,B_s)\end{matrix}\right.\right] \tag{3.5.1}$$

$$= 2^{\rho+\sigma} \sum_{N=0}^{\infty} \sum_{\mu=0}^{N} \frac{(2N-m+n-1)\,\Gamma(N-m+1)}{N!\,\mu!\,\Gamma(1+n+N)}$$

$$\times \frac{\Gamma(1+n-m+N+\mu)\,(-N)_{\mu}}{\Gamma(1-m+\mu)} P^{m,n}_{N-(m-n)/2}(x)$$

$$\times H^{p,q+2}_{r+2,s+1}\left[2^{\nu+\delta}z \left| \begin{matrix} (-n-\sigma,\delta),(m-\rho-\mu,\nu),(a_r,A_r) \\ (b_s,B_s),(-1-\rho-\sigma-\mu+m-n,\gamma+\delta) \end{matrix} \right.\right]$$

This formula (3.5.1) holds under the following conditions :

(i) $\delta > 0$, $\nu \geqslant 0$, (or $\delta \geqslant 0$, $\nu > 0$).

(ii) p, q, r and s are positive integers such that $1 \leqslant p \leqslant s, 0 \leqslant q \leqslant r$.

(iii) $\sum_{1}^{s} B_j - \sum_{1}^{r} A_j > 0$, when $z \neq 0$ and if $\sum_{1}^{s} B_j - \sum_{1}^{r} A_j = 0$,

then

$$0 < |z| < \prod_{j=1}^{r} (A_j)^{-A_j} \prod_{j=1}^{s} (B_j)^{B_j} .$$

(iv) $A_j(b_h + \xi) \neq B_h(a_j - \eta - 1)$,

$\delta(b_h + \xi) \neq -B_h(n + \sigma + \eta + 1)$,

$\nu(b_h + \xi) \neq B_h(m - \rho - \eta - 1)$,

$(\xi, \eta = 0, 1, \ldots; h = 1, \ldots, p$ and $j = 1, \ldots, q)$.

(v) $|\arg z| < \pi\lambda/2$ where,

$$\lambda = \sum_{1}^{q} A_j - \sum_{q+1}^{r} A_j + \sum_{1}^{p} B_j - \sum_{p+1}^{s} B_j > 0.$$

(vi) $R(n) > -1$, $R(m) < 1$, $R(\rho - m + \nu b_j/B_j) > -1$

$R(\sigma + n + \delta\, b_j/B_j) > -1 \quad (j = 1, \ldots, p)$.

REMARK. The orthogonality property of generalized Legendre's associated functions is given by the relation (Meulenbeld and Robin, 1961) as follows:

$$\int_{-1}^{1} P^{m,n}_{k-(m-n)/2}(x)\; P^{m,n}_{l-(m-n)/2}(x)\,dx \tag{3.5.2}$$

$= 0$ if $k \neq l$

$$= \frac{2^{n-m+1}\,k!\,\Gamma(k+n+1)}{\Gamma(2k-m+n+1)\,\Gamma(k-m+1)\,\Gamma(k-m+n+1)} \text{ if } k = l,$$

where $R(m) < 1$ and $R(n) > -1$.

REMARK 2. For a detailed discussion on generalized Legendre's associated functions, see the original papers by Kuipers and Meulenbeld (1957), Meulenbeld (1958) and Meulenbeld and Robin (1961).

3.6 AN EXPANSION FORMULA FOR THE PRODUCTS OF TWO GENERALIZED HYPERGEOMETRIC FUNCTIONS

On using the integral (2.9.4) and the orthogonality property of the associated Legendre functions, it can be easily shown that

$$w^{\rho-1} {}_uF_v\left(\begin{matrix}\alpha_u\\ \beta_v\end{matrix}; cw^d\right) H_{P,Q}^{K,L}\left[zw^m \middle| \begin{matrix}(a_P, A_P)\\ (b_Q, B_Q)\end{matrix}\right] \tag{3.6.1}$$

$$= \pi\, 2^{\mu-1} \sum_{n,r=o}^{\infty} \frac{\prod_{j=1}^{u} (\alpha_j; r)\,(2n+1)\,(n-\mu)!\, c^r}{\prod_{j=1}^{v} (\beta_j; r)\,(n+\mu)!\,(r)!}$$

$$\times \frac{P_n^{\mu}\{\pm(1-w)^{1/2}\}}{\Gamma\left(\frac{2+n-\mu}{2}\right)\Gamma\left(\frac{1-n-\mu}{2}\right)}$$

$$\times H_{P+2,Q+2}^{K,L+2}\left[z \middle| \begin{matrix}(1-\rho-rd\pm\mu/2, m), (a_P, A_P)\\ (b_Q, B_Q), (1-\rho-rd+n/2, m), (-\rho-rd-n/2, m)\end{matrix}\right]$$

which is valid under the same conditions stated with (2.9.4) with $\rho \geqslant 1$.

For $c = 0$, (3.6.1) reduces to

$$w^{\rho-1} H_{P,Q}^{K,L}\left[zw^m \middle| \begin{matrix}(a_P, A_P)\\ (b_Q, B_Q)\end{matrix}\right] \tag{3.6.2}$$

$$= \pi\, 2^{\mu-1} \sum_{n=o}^{\infty} \frac{(2n+1)\,(n-\mu)!\, P_n^{\mu}\{\pm(1-w)^{1/2}\}}{(n+\mu)!\,\Gamma\left[\frac{1}{2}(2+n-\mu)\right]\Gamma\left(\frac{1-n-\mu}{2}\right)}$$

$$\times H_{P+2,Q+2}^{K,L+2}\left[z \middle| \begin{matrix}(1-\rho\pm\mu/2, m), (a_P, A_P)\\ (b_Q, B_Q), (1-\rho+n/2, m), (-\rho-n/2, m)\end{matrix}\right]$$

The results (3.6.1) and (3.6.2) are given by Singh and Varma (1972).

3.7 REPRESENTATION OF AN *H*-FUNCTION IN COMPUTABLE FORM

CASE I. When the poles of $\prod_{j=1}^{m} \Gamma(b_j - B_j s)$ are simple, that is, where

$$B_h(b_j+\lambda) \neq B_j(b_h+\nu)$$

for $j \neq h$; $j, h = 1, \ldots, m$; $\lambda, \nu = 0, 1, 2, \ldots$,; then we obtain the following expansion for the H-function.

$$H_{p,q}^{m,n}(z) \tag{3.7.1}$$

$$= \sum_{h=1}^{m} \sum_{\nu=0}^{\infty} \frac{\prod\limits_{j=1,\, j\neq h}^{m} \Gamma\{(b_j - B_j(b_h+\nu)/B_h)\} \prod\limits_{j=1}^{n} \Gamma\left\{1 - a_j + A_j \dfrac{(b_h+\nu)}{B_j}\right\}}{\prod\limits_{j=m+1}^{q} \Gamma\left\{1 - b_j + B_j \dfrac{(b_h+\nu)}{B_h}\right\} \prod\limits_{j=n+1}^{p} \Gamma\left\{a_j - A_j \dfrac{(b_h+\nu)}{B_h}\right\}}$$
$$\times \frac{(-1)^\nu z^{(b_h+\nu)/B_h}}{(\nu)!\, B_h}$$

which exists for all $z \neq 0$ if $\mu > 0$ and for $0 < |z| < \beta^{-1}$ if $\mu = 0$ where

$$\mu = \sum_{j=1}^{q} B_j - \sum_{j=1}^{p} A_j$$

and

$$\beta = \prod_{j=1}^{p} A_j^{A_j} \prod_{j=1}^{q} B_j^{-B_j}.$$

This result is given by Braaksma (1964).

CASE II. When the poles of $\prod\limits_{j=1}^{n} \Gamma(1 - a_j + A_j s)$ are simple, that is, when

$$A_h(1 - a_j + \nu) \neq A_j(1 - a_j + \lambda),$$

$j \neq h$; $j, h = 1, \ldots, n$; $\lambda, \nu = 0, 1, \ldots$; we have the following expansion for the H-function.

$$H_{p,q}^{m,n}(z) \tag{3.7.2}$$

$$= \sum_{h=1}^{n} \sum_{\nu=0}^{\infty} \frac{\prod\limits_{j=1,\, j\neq h}^{n} \Gamma[1 - a_j - A_j(1 - a_h + \nu)/A_h]}{\prod\limits_{j=n+1}^{p} \Gamma[a_j + A_j(1 - a_h + \nu)/A_h]}$$
$$\times \frac{\prod\limits_{j=1}^{m} \Gamma[b_j + B_j(1 - a_h + \nu)/A_h]}{\prod\limits_{j=m+1}^{q} \Gamma[1 - b_j - B_j(1 - a_h + \nu)/A_h]} \frac{(-1)^\nu}{\nu!\, A_h} \left(\frac{1}{z}\right)^{(1-a_h+\nu)/A_h}$$

which exists for all $z \neq 0$ if $\mu < 0$ and for $|z| > 1/\beta$ if $\mu = 0$ where μ and β are defined in (3.7.1).

NOTE. Formula (3.7.2) is available from (3.7.1) by making the following changes in it. In (3.7.1) interchange

$$m \sim n,\ q \sim p$$
$$b_j \sim 1 - a_j$$
$$B_j \sim A_j \text{ and } z \sim 1/z.$$

GENERAL CASE. Here we take the H-function in the following form:

$$H(z) = \frac{1}{2\pi i}\int_L \chi(s)\, z^s\, ds, \tag{3.7.3.}$$

where

$$\chi(s) = \frac{\prod_{j=1}^{m}\Gamma(b_j - B_j s)\ \prod_{j=1}^{n}\Gamma(1 - a_j + A_j s)}{\prod_{j=m+1}^{q}\Gamma(1 - b_j + B_j s)\ \prod_{j=n+1}^{p}\Gamma(a_j + A_j s)} \tag{3.7.4}$$

General Expansion in Case I. It is general in the sense that the poles of

$$\prod_{j=1}^{m}\Gamma(b_j - B_j s)$$

and not assumed to be simple.

The expansion will be written down with the help of the following equations.

$$s = \frac{b_j + \nu}{B_j},\ \nu = 0, 1, \ldots \tag{3.7.5}$$

There may exist a pair of values (ν_1, λ_1) such that

$$\frac{b_j + \nu_1}{B_j} = \frac{b_h + \lambda_1}{B_h},\ h \neq j. \tag{3.7.6}$$

Then evidently the point

$$s = \frac{b_j + \nu_1}{B_j}$$

is a pole of order two if the point does not coincide with the pole of any other gamma of $\chi(s)$.

If the point coincides with the poles coming from $(r - 1)$ other gammas of the set $\Gamma(b_j - B_j s)$, $j = 1, \ldots, m$, then the point gives a pole of order r. For convenience we will assume that no singularity of any gamma of the denominator of $h(s)$ coincides with any singularity of any gamma of the set $\Gamma(b_j - B_j s)$, $j = 1, \ldots, m$.

For a fixed j consider the equations

$$\frac{b_1 + \nu^{(j1)}_{j_1 \ldots j_m}}{B_1} = \frac{b_2 + \nu^{(j2)}_{j_1 \ldots j_m}}{B_2} = \ldots = \frac{b_m + \nu^{(jm)}_{j_1 \ldots j_m}}{B_m} \tag{3.7.7}$$

The following convention is used in (3.7.7). For a fixed j, $j_r = 0$ or 1 for $r = 1, 2, \ldots, m$. If $j_r = 0$ then $(b_r + \nu^{(jr)}_{j_1\ldots j_m})/B_r$ is to be excluded from the equations in (3.7.7). Here $\nu^{(jr)}_{j_1\ldots j_m}$ is a value of ν in (3.7.5). Evidently the possible values are $0, 1, 2, \ldots$. Also $\nu^{(jr)}_{j_1\ldots j_m}$ denotes the number corresponding to $\nu^{(jj)}_{j_1\ldots j_m}$ when the equation

$$\frac{b_j + \nu^{(jj)}_{j_1\ldots j_m}}{B_j} = \frac{b_r + \nu^{(jr)}_{j\ldots j_m}}{B_r} \tag{3.7.8}$$

is satisfied by some values of $\nu^{(jj)}_{j_1\ldots j_m}$ and $\nu^{(jr)}_{j_1\ldots j_m}$. Thus $\nu^{(jr)}_{j_1\ldots j_m}$ may or may not exist. Under these conditions

$$0 \leqslant j_1 + \ldots + j_m \leqslant m$$

for every fixed j and $j_1 + \ldots + j_m$ denotes the order of the pole at

$$s = \frac{(b_j + \nu^{(jj)}_{j_1\ldots j_m})}{B_j}. \tag{3.7.9}$$

For example if $j_1 + \ldots + j_m = r$, then in (3.7.7) there will be r elements if

$$\frac{(b_k + \nu^{(jk)}_{j_1\ldots j_m})}{B_k},\ k = 1, 2, \ldots, m$$

are called elements in (3.7.7). In the above notation a pole of order r will be considered r times. In order to avoid duplication we will always assume that

$$j_1 = j_2 = \ldots = j_{j-1} = 0 \tag{3.7.10}$$

while considering the points corresponding to (3.7.5).
If $j_1 + \ldots + j_m = 0$, then the corresponding point is not a pole.
If $j_r = 0$ for $r = 1, \ldots, m$; $r \neq j$ then

$$s = \frac{(b_j + \nu^{(jj)}_{j_1\ldots j_m})}{B_j}$$

gives a simple pole.

Let

$$S^{(jj)}_{j_1\ldots j_m} = \{\nu^{(jj)}_{j_1\ldots j_m}\} \tag{3.7.11}$$

be the set of all values $\nu^{(jj)}_{j_1\ldots j_m}$ takes for given $j_1, \ldots, j_m$. Now the H-function can be written as

$$H(z) = \sum_{j=1}^{m} \sum_{S^{(jj)}_{j_1 \dots j_m}} R_j \tag{3.7.12}$$

where R_j is the residue of $\chi(s)\, z^s$ at the pole

$$s = \frac{(b_j + \nu^{(jj)}_{j_1 \dots j_m})}{B_j} \text{ and}$$

$\sum_{S^{(jj)}_{j_1 \dots j_m}}$ denotes the summation over all sets $S^{(jj)}_{j_1 \dots j_m}$.

$$R_j = \frac{(j_1 + \dots + j_m)\, z^{(b_j + \nu^{(jj)}_{j_1 \dots j_m})/B_j}}{(j_1 + \dots + j_m)!} \sum_{r=0}^{j_1+\dots+j_m-1} \binom{j_1 + \dots + j_m - 1}{r}$$

$$\times (-\log z)^{j_1+\dots+j_m-1-r} \left[\sum_{r_1=0}^{r-1} \binom{r-1}{r_1} c_j^{(r-1-r_1)} \right.$$

$$\left. \times \sum_{r_2=0}^{r_1-1} \binom{r_1 - 1}{r_2} c_j^{(r_1-r-r_2)} \dots \right] D_j; \tag{3.7.13}$$

$$D_j = \left\{ \prod_{h=1}^{m} \Gamma\,[b_h - B_h\,(b_j + \nu^{(jj)}_{j_1 \dots j_m})/B_j + j_h\,(\nu^{(jh)}_{j_1 \dots j_m} + 1)] \right.$$

$$\times \prod_{h=1}^{n} \Gamma[1 - a_h + A_h\,(b_j + \nu^{(jj)}_{j_1 \dots j_m})/B_j] \Big\} \Big/ \Big\{ \prod_{h=1}^{m} B_h^{j_h}\,(-1)^{j_h}\,(\nu^{(jh)}_{j_1 \dots j_m})$$

$$\times (\nu^{(jh)}_{j_1 \dots j_m}!)^{j_h}] \prod_{h=m+1}^{q} \Gamma[1 - b_h + B_h\,(b_j + \nu^{(jj)}_{j_1 \dots j_m})/B_j]$$

$$\left. \times \prod_{h=n+1}^{p} \Gamma\,[a_h - A_h\,(b_j + \nu^{(jj)}_{j_1 \dots j_m})/B_j] \right\};$$

$$C_j^{(0)} = \sum_{h=1}^{m} B_h \psi\,[b_h - B_h\,(b_j + \nu_{j j_1 \dots j_m}^{(jj)})/B_j + j_h\,(\nu^{(jh)}_{j_1 \dots j_m} + 1)]$$

$$+ \sum_{h=1}^{m} B_h j_h \left[\frac{1}{1} + \frac{1}{2} + \dots + \frac{1}{\nu^{(jh)}_{j_1 \dots j_m}} \right]$$

$$- \sum_{h=1}^{n} A_h \psi\,[1 - a_h + A_h\,(b_j + \nu^{(jj)}_{j_1 \dots j_m})/B_j]$$

$$+ \sum_{h=m+1}^{q} B_h \psi\,[\;- b_h + B_h\,(b_j + \nu^{(jj)}_{j_1 \dots j_m})/B_j]$$

$$- \prod_{h=n+1}^{p} A_h \psi\,[a_j - A_h\,(b_j + \nu^{(jj)}_{j_1 \dots j_m})/B_j]; \tag{3.7.14}$$

$$C_j^{(t)},\ t \geqslant 1, = (-1)^{t+1}\, t!\left\{ \sum_{h=1}^{m} B_h^{t+1}\, \rho[t+1, b_h - B_h(b_j + \nu^{(jj)}_{j_1 \ldots j_m})/B_j + j_h(\nu^{(jj)}_{j_1 \ldots j_m} + 1)] \right. \tag{3.7.15}$$

$$+ \sum_{h=1}^{m} B_h^{t+1} j_h \left[\frac{1}{(-1)^{t+1}} + \frac{1}{(-2)^{t+1}} + \ldots + \frac{1}{(-\nu^{(jh)}_{j_1 : j_m})^{t+1}}\right]$$

$$+ \sum_{h=1}^{n} (-A_h)^{t+1} \rho[t+1, 1 - a_h + A_h(b_j + \nu^{(jj)}_{j_1 \ldots j_m})/B_j]$$

$$- \sum_{h=m+1}^{q} (-B_h)^{t+1}\, \rho[t+1, 1 - b_h + B_h(b_j + \nu^{(jj)}_{j_1 \ldots j_m})/B_j]$$

$$\left. - \sum_{h=n+1}^{p} (A_h)^{t+1} \rho[t+1, a_h - A_h(b_j + \nu^{(jj)}_{j_1 \ldots j_m})/B_j] \right\}$$

The psi function $\psi(\cdot)$ and the generalized zeta function $\rho(\cdot,\cdot)$ are defined as follows:

$$\psi(z) = \frac{d}{dz}\log\Gamma(z) = -\gamma + (z-1)\sum_{n=0}^{\infty}\frac{1}{(n+1)(z+n)}, \tag{3.7.16}$$

where γ is the Euler's constant; $\gamma = 0.577\ldots$.

$$\rho(s, \nu) = \sum_{n=0}^{\infty}\frac{1}{(n+\nu)^s},\quad R(s) > 1,\ \nu \neq 0, -1, -2, \ldots; \tag{3.7.17}$$

where $R(\cdot)$ denotes the real part of $(\cdot)$.

NOTE. Case II is available from case I by making the following changes. Interchange m and n, p and q, b_j and $1 - a_j$, B_j and A_j. For further details on the expansion see Mathai (1973b).

EXERCISES

3.1 Show that

(i) $$z^\lambda = \frac{\prod_{j=m+1}^{q}\Gamma(1 - b_j - B_j\lambda)\ \prod_{j=n+1}^{p}\Gamma(a_j + A_j\lambda)}{\prod_{j=1}^{m}\Gamma(b_j + B_j\lambda)\ \prod_{j=1}^{n}\Gamma(1 - a_j - A_j\lambda)}$$

$$\times \sum_{r=0}^{\infty}\frac{2(\lambda + r)\,\Gamma(2\lambda + r)}{r!}$$

$$\times H^{m,n+1}_{p+2,q}\left[\frac{1}{z}\middle|\begin{matrix}(1-\lambda-r, 1), (a_p, A_p), (1+\lambda+r, 1)\\ (b_q, B_q)\end{matrix}\right]$$

where $R(1 - b_j - B_j\lambda) > 0$ $(j = m+1, \ldots, q)$ and $R(a_j + \lambda A_j) > 0$ $(j = n+1, \ldots, p)$.

(ii) $$\sum_{r=0}^{\infty} \frac{(\nu)_r}{(\sigma)_r} H_{p,q}^{m,n}\left[z \middle| \begin{matrix}(a_1, 1), (a_2 + A_2 r, A_2), \ldots, (a_p + A_p r, A_p) \\ (b_1 + B_1 r, B_1), \ldots, (b_q + B_q r, B_q)\end{matrix}\right] \frac{z^{-r}}{r!}$$

$$= \frac{\Gamma(\sigma)}{\Gamma(\sigma - \nu)}$$

$$\times H_{p+1,q+1}^{m+1,n}\left[z \middle| \begin{matrix}(a_1, 1), (a_2, A_2), \ldots, (a_p, A_p), (a_1 + \sigma - 1, 1) \\ (a_1 + \sigma - \nu - 1, 1), (b_q, B_q)\end{matrix}\right]$$

where $R(a_1 - \nu + \sigma - 1) > 0$.

(iii) $$\sum_{r=0}^{\infty} \frac{z^{-r}}{r!\,\Gamma(\lambda + r)}$$

$$H_{p,q}^{m,n}\left[z \middle| \begin{matrix}(a_1, 1), (a_2, 1), \ldots, (a_3 + rA_3, A_3), \ldots, (a_p + rA_p, A_p) \\ (b_q + rB_q, B_q)\end{matrix}\right]$$

$$= H_{p+2,q+1}^{m+1,n}$$

$$\left[z \middle| \begin{matrix}(a_1, 1), (a_2, 1), (a_3, A_3), \ldots, (a_p, A_p), (a_1 + \lambda - 1, 1), (a_2 + \lambda - 1, 1) \\ (a_1 + a_2 + \lambda - 2, 2), (b_q, B_q)\end{matrix}\right]$$

where $R(a_1 + a_2 + \lambda - 2) > 0$

(Srivastava, H.M. and Daoust, M.C., 1969a)

3.2 Derive the following finite and infinite series expansions.

(i) $$\sum_{k=0}^{r} \frac{(-r)_k}{k!}$$

$$H_{p+2,q+2}^{m+2,n}\left[\lambda z \middle| \begin{matrix}(\alpha_p, \gamma_p), (c + k, h), (1 - b - k, h) \\ (a + k, h), (c - a - b + r - k, h), (\beta_q, \delta_q)\end{matrix}\right]$$

$$= (c - a)_r$$

$$H_{p+2,q+2}^{m+3,n}\left[\lambda z \middle| \begin{matrix}(\alpha_p, \gamma_p), (c + r, h), (1 - b, h), (c - b, 2h) \\ (a, h), (c - a - b, h), (c - b + r, 2h), (\beta_q, \delta_q)\end{matrix}\right]$$

(ii) $$\sum_{k=0}^{\infty} \frac{(a)_k}{k!}$$

$$H_{p+2,q+2}^{m+2,n}\left[\lambda z \middle| \begin{matrix}(\alpha_p, \gamma_p), (a + c + k, h), (c - k, h) \\ (b + k, h), (b - a - k, h), (\beta_q, \delta_q)\end{matrix}\right]$$

$$= \frac{\Gamma(1 + a/2)}{\Gamma(1 + a)} \frac{\Gamma(a/2 - b - c)}{\Gamma(a - b + c)}$$

$$H_{p+3,q+3}^{m+2,n+1}\left[\lambda z \middle| \begin{matrix}(b - a, h), (\alpha_p, \gamma_p), (c, h), (a/2 + c, h) \\ (b, h), (b - a, h), (\beta_q, \delta_q), (b - a/2, h)\end{matrix}\right]$$

(iii) $$\sum_{k=0}^{\infty} \frac{(a)_k\,(1 - a)_k}{k!\,(1 + 2c - f)_k} H_{p+1,q+1}^{m+1,n}\left[\lambda z \middle| \begin{matrix}(\alpha_p, \gamma_p), (f + k, 2h) \\ (c + k, h), (\beta_q, \delta_q)\end{matrix}\right]$$

$$= \frac{\pi\,\Gamma(1+2c-f)}{\Gamma\left(\frac{1}{2}+\frac{a}{2}+c-\frac{f}{2}\right)\Gamma\left(1-\frac{a}{2}+c-\frac{f}{2}\right)}$$

$$\times H^{m+1,n}_{p+2,q+1}\left[4^h\lambda z\,\middle|\,\begin{matrix}(\alpha_p,\gamma_p),\left(\frac{a}{2}+f,h\right),\left(\frac{1}{2}-\frac{a}{2}+\frac{f}{2},\ h\right)\\(c,h),(\beta_q,\delta_q)\end{matrix}\right]$$

(iv) $$\sum_{k=0}^{\infty}\frac{(-1)^k\,(a)_k\,(1+a/2)_k}{k!\,(a/2)_k}$$

$$H^{m+2,n}_{p+2,q+2}\left[\lambda z\,\middle|\,\begin{matrix}(\alpha_p,\gamma_p),(1+a-c+k,h),(1-c-k,h)\\(b+k,h),(b-a-k,h),(\beta_q,\delta_q)\end{matrix}\right]$$

$$=\frac{1}{\Gamma(1+a)\,\Gamma(1+a-b-c)}$$

$$H^{m+2,n+1}_{p+2,q+2}\left[\lambda z\,\middle|\,\begin{matrix}(b-a,h),(\alpha_p,\gamma_p),(1-c,h)\\(b,h),(b-a,h),(\beta_q,\delta_q)\end{matrix}\right]$$

(v) $$\sum_{k=0}^{\infty}\frac{(a)_k}{k!}$$

$$H^{m+2,n}_{p+2,q+2}\left[\begin{matrix}\lambda z\,(\alpha_p,\gamma_p),\left(\frac{1}{2}+\frac{a}{2}+\frac{b}{2}+k,h\right),(1-c-k,h)\\(b+k,2h),(1-2c-k,2h),(\beta_q,\delta_q)\end{matrix}\right]$$

$$=\frac{\Gamma(\frac{1}{2})}{\Gamma\left(\frac{1+a}{2}\right)}\,H^{m+2,n+2}_{p+4,q+4}$$

$$\left[\lambda z\,\middle|\,\begin{matrix}\left(\frac{1}{2}-c,h\right),\left(\frac{1}{2}+\frac{a}{2}+\frac{b}{2}-c,2h\right),(\alpha_p,\gamma_p),\left(\frac{1+b}{2},h\right),(1-c,h)\\(b,2h),(1-2c,2h),(\beta_q,\delta_q),\left(\frac{1}{2}+\frac{a}{2}-c,h\right),\left(\frac{1}{2}+\frac{b}{2}-c,2h\right)\end{matrix}\right]$$

(vi) $$\sum_{r=0}^{\infty}\frac{(-r)_k\,(b+r)_k}{k!\left(\frac{1}{2}+\frac{b}{2}\right)_k\left(1+\frac{b}{2}\right)_k}$$

$$H^{m+2,n}_{p+1,q+2}\left[\lambda z\,\middle|\,\begin{matrix}(\alpha_p,\gamma_p),(1+a+k,2h)\\\left(\frac{1}{2}+\frac{a}{2}+k,h\right),\left(1+\frac{a}{2}+k,h\right),(\beta_q,\delta_q)\end{matrix}\right]$$

$$=\frac{b}{(b+2r)\,(b)_r}$$

$$H^{m+2,n+1}_{p+2,q+3}\left[\lambda z\,\middle|\,\begin{matrix}(2+a-b-r,2h),(\alpha_p,\gamma_p),(1+a,2h)\\\left(\frac{1}{2}+\frac{a}{2},h\right),\left(1+\frac{a}{2},h\right),(\beta_q,\delta_q),(2+a-b,2h)\end{matrix}\right]$$

(Jain, R.N. 1969)

3.3 If ρ is an integer, show that

$$\int_0^\pi (\sin\phi)^{-2\rho}\cos(2u\phi)\, H_{p,q}^{m,n}\left[z(\sin\phi)^{-2h}\,\middle|\,\begin{matrix}(a_p, A_p)\\(b_q, B_q)\end{matrix}\right]d\phi$$

$$= \sqrt{\pi}\, H_{p+2,q+2}^{m+1,n+1}\left[z\,\middle|\,\begin{matrix}(1-\rho-u, h), (a_p, A_p), (1-\rho+u, h)\\(\tfrac{1}{2}-\rho, h), (b_q, B_q), (1-\rho, h)\end{matrix}\right]$$

where

$$h>0,\ \mu=\sum_1^p A_j-\sum_1^q B_j\leqslant 0,\ \lambda=\sum_1^n A_j-\sum_{n+1}^p A_j+\sum_1^m B_j-\sum_{m+1}^q B_j>0,$$

$$|\arg z|<\pi\lambda/2,\ (u=0, 1, 2, \ldots).$$

Hence derive the recurrence relation

$$H_{p+2,q+2}^{m+1,n+1}\left[z\,\middle|\,\begin{matrix}(1-\rho-u, h), (a_p, A_p), (1-\rho+u, h)\\(\tfrac{1}{2}-\rho, h), (b_q, B_q), (1-\rho, h)\end{matrix}\right]$$

$$-H_{p+2,q+2}^{m+1,n+1}\left[z\,\middle|\,\begin{matrix}(1-\rho-u, h), (a_p, A_p), (2-\rho+u, h)\\(\tfrac{1}{2}-\rho, h), (b_q, B_q), (1-\rho, h)\end{matrix}\right]$$

$$=2\,H_{p+2,q+2}^{m+1,n+1}\left[z\,\middle|\,\begin{matrix}(1-\rho-u, h), (a_p, A_p), (2-\rho+u, h)\\(3/2-\rho, h), (b_q, B_q), (1-\rho, h)\end{matrix}\right]$$

Hint. Use the identity

$$\sin((2\phi+1)\theta)\sin\theta=\tfrac{1}{2}[\cos(2\phi\theta)-\cos((2\phi+2)\theta)]$$

(Bajpai, 1969b)

3.4 Show that

$$(1-x)^\sigma\, H_{p,q}^{m,n}\left[z(1-x)^k\,\middle|\,\begin{matrix}(a_p, A_p)\\(b_q, B_q)\end{matrix}\right]$$

$$=2^\sigma\sum_{r=0}^\infty\frac{(-1)^r(\alpha+\beta+2r+1)\,\Gamma(\alpha+\beta+r+1)}{\Gamma(\alpha+r+1)}P_r^{(\alpha,\beta)}(x)$$

$$\times H_{p+2,q+2}^{m,n+2}\left[2^k z\,\middle|\,\begin{matrix}(-\alpha-\sigma, k), (-\sigma, k), (a_p, A_p)\\(b_q, B_q), (-\sigma+r, k), (-1-\alpha-\beta-\sigma-r, k)\end{matrix}\right]$$

$$=2^\sigma\sum_{r=0}^\infty\frac{(\alpha+\beta+2r+1)\,\Gamma(\alpha+\beta+r+1)}{\Gamma(\alpha+r+1)}P_r^{(\alpha,\beta)}(x)$$

$$\times H_{p+2,q+2}^{m+1,n+1}\left[2^h z\,\middle|\,\begin{matrix}(-\alpha-\sigma, k), (a_p, A_p), (-\sigma, k)\\(-\sigma+r, k), (b_q, B_q), (-1-\alpha-\beta-\sigma-r, k)\end{matrix}\right]$$

where $\mu\leqslant 1$, $\lambda>0$, $|\arg z|<\pi\lambda/2$, $R(\alpha)>-1$, $R(\beta)>-1$, $\sigma\geqslant 0$, $-1<x<1$, $R(\sigma+\alpha+k\min b_j/B_j)>0$ $(j=1,\ldots,m)$ and λ and μ are defined in exercise 3.3.

(Bajpai, 1969a)

3.5 Show that

$$x^{\rho}\ H_{p,q}^{m,n}\left[zx^h \middle| \begin{matrix}(a_p, A_p)\\(b_q, B_q)\end{matrix}\right]$$

$$= \sum_{r=0}^{\infty} \frac{(-1)^r}{\Gamma(\alpha+r+1)} L_r^{(\alpha)}(x)$$

$$\times H_{p+2,q+1}^{m,n+2}\left[z \middle| \begin{matrix}(-\beta-\alpha, h), (-\beta, h), (a_p, A_p)\\(b_q, B_q), (r-\beta, h)\end{matrix}\right]$$

where $h > 0$, $|\arg z| < \pi\lambda/2$, $R(\beta + h \min b_j/B_j) > -1$ $(j = 1, \ldots, m)$, $R(\alpha) > -1$ and $\mu \leqslant 1$.

(Bajpai, 1969)

3.6 Prove the following expansions

$$x^{\alpha-1/2}\ H_{p,q}^{m,n}\left[zx^h \middle| \begin{matrix}(a_p, A_p)\\(b_q, B_q)\end{matrix}\right]$$

$$= \frac{2}{\sqrt{\pi}} \sum_{u=1}^{\infty} T_u(2x-1)$$

$$\times H_{p+2,q+2}^{m,n+2}\left[z \middle| \begin{matrix}(1-\alpha, h), (\tfrac{1}{2}-\alpha, h), (a_p, A_p)\\(b_q, B_q),\ (\tfrac{1}{2}-\alpha+u, h),\ (\tfrac{1}{2}-\alpha-u, h)\end{matrix}\right]$$

and

$$x^{\alpha-1/2}\ H_{p,q}^{m,n}\left[zx^{-h} \middle| \begin{matrix}(a_p, A_p)\\(b_q, B_q)\end{matrix}\right]$$

$$= \frac{2}{\sqrt{\pi}} \sum_{u=1}^{\infty} T_u(2x-1)\ H_{p+2,q+2}^{m+2,n}\left[z \middle| \begin{matrix}(a_p, A_p), (\alpha \pm u + \tfrac{1}{2}, h)\\(d, h), (d+\tfrac{1}{2}, h), (b_q, B_q)\end{matrix}\right]$$

where h is a positive number, $\mu \leqslant 1$, $\lambda > 0$, $|\arg z| < \pi\lambda/2$,

$$R[\alpha + h \min b_j/B_j] > -\tfrac{1}{2}\ (j = 1, \ldots, m),$$

$$R\left[\alpha - h\frac{(a_j-1)}{A_j}\right] < -\tfrac{1}{2}\ (j = 1, \ldots, n).$$

(Dahiya, 1971)

REMARK. For the expansions of products of two H-functions in series of Hermite polynomials and generalized H-function, see the work of Saxena and Modi (1974).

3.7 Prove that

$$t^{\omega}\ H_{E,(A:C),F,(B:D)}^{L,N,N_1,M,M_1}\left[\begin{matrix}xt^{\sigma}\\yt^{\sigma}\end{matrix}\right]$$

$$= \sum_{r=1}^{\infty} \frac{(-1)^r}{\Gamma(\alpha + r + 1)} L_r^{(\alpha)}(t)$$

$$\times H_{E+2,(A:C),F+1,(B:D)}^{L+2,N,N_1,M,M_1} \left[\begin{matrix} x \\ y \end{matrix} \middle| \chi_1 \right]$$

where χ_1 denotes the parameters

$$\left[\begin{matrix} (e_E, \theta_E), (1 + \omega + \alpha, \sigma), (1 + \omega, \sigma) \\ (a_A, \alpha_A); (c_C, \gamma_C) \\ (f_F, \phi_F); (1 + \omega - r, \sigma) \\ (b_B, \beta_B); (d_D, \delta_D) \end{matrix} \right]$$

where σ is a positive number, $\rho_j > 0$ $(j = 1, 2)$, $|\arg x| < \pi\rho_1/2$ and $|\arg y| < \pi\rho_2/2$, $R(\omega + \sigma\, b_j/B_j + \sigma d_h/\delta_h) > -1$ $(j = 1, \ldots, M;\ h = 1, 2, \ldots, M_1)$; $R(\alpha) > -1$.

Hence deduce that

$$t^{\omega}\ H_{A,B}^{M,N} \left[xt^{\sigma} \middle| \begin{matrix} (a_A, \alpha_A) \\ (b_B, \beta_B) \end{matrix} \right] H_{C,D}^{M_1,N_1} \left[yt^{\sigma} \middle| \begin{matrix} (c_C, \gamma_C) \\ (d_D, \delta_D) \end{matrix} \right]$$

$$= \sum_{r=0}^{\infty} \frac{(-1)^r}{\Gamma(\alpha + r + 1)} L_r^{(\alpha)}(t)$$

$$\times H_{2,(A:C),1,(B:D)}^{2,N,N_1,M,M_1} \left[\begin{matrix} x \\ \\ y \end{matrix} \middle| \begin{matrix} (1 + \omega + \alpha, \sigma), (1 + \omega, \sigma) \\ (a_A, \alpha_A); (c_C, \gamma_C) \\ (1 + \omega - r, \sigma) \\ (b_B, \beta_B); (d_D, \delta_D) \end{matrix} \right]$$

(Saxena, 1971b)

3.8 Show that

$$\int_0^{\infty} e^{-t}\, t^{\gamma-1}\, Y_n(1, a; t)\, H_{p,q}^{m,n} \left[bt \middle| \begin{matrix} (a_p, A_p) \\ (b_q, B_q) \end{matrix} \right] dt$$

$$= H_{p+2,q+1}^{m,n+2} \left[b \middle| \begin{matrix} (1+n-\gamma, 1), (2-\gamma-n-a, 1), (a_p, A_p) \\ (b_q, B_q), (2-\gamma-a, 1) \end{matrix} \right]$$

where $R(\gamma + \min b_j/B_j) > 0$ $(j = 1, \ldots, m)$; $|\arg b| < \lambda\pi/2$

$$\lambda = \sum_{j=1}^{n} A_j - \sum_{j=n+1}^{p} A_j + \sum_{j=1}^{m} B_j - \sum_{j=m+1}^{q} B_j > 0 \text{ and } \mu \leqslant 0.$$

Hence, using the orthogonality property of the Bessel polynomials, prove that

$$t^{\omega} H_{p,q}^{m,n}\left[bt \left| \begin{matrix}(a_p, A_p)\\(b_q, B_q)\end{matrix}\right.\right]$$

$$= \sum_{r=0}^{\infty} \frac{1}{r!\,\Gamma(2-a-r)} Y_r(1, a, t)$$

$$\times H_{p+2,q+1}^{m,n+2}\left[b \left| \begin{matrix}(a+r-\omega-1, 1), (-r-\omega, 1), (a_p, A_p)\\(b_q, B_q), (-\omega, 1)\end{matrix}\right.\right]$$

where

$$R(\omega - a + \min b_j/B_j) > 0 \; (j = 1, \ldots, m); \; |\arg b| < \lambda\pi/2 \text{ and } \mu \leqslant 0.$$

NOTE. Orthogonality property of the Bessel polynomials is given in the Appendix.

CHAPTER 4

Applications in Statistics and Other Disciplines

In this chapter we discuss the applications of the H-function in deriving the exact distributions of products, ratios and linear combinations of certain types of random variables. We will point out situations where the H-function plays a vital role in getting the distributions of the likelihood ratio test criteria in certain multivariate statistical problems. We will discuss the role of H-functions in the characterizations of probability laws, generalized probability laws and in the structural analysis of probability models. Also the possible application of an H-function in some problems in physical sciences will be indicated in this chapter.

4.1 PRODUCTS AND RATIOS OF GENERALIZED GAMMA RANDOM VARIABLES

A random variable X is called a generalized gamma variate if it has the density function,

$$f(x)=\begin{cases} \dfrac{\beta a^{\alpha/\beta}}{\Gamma\left(\dfrac{\alpha}{\beta}\right)} x^{\alpha-1} e^{-ax^{\beta}}, \; x>0, a>0, \alpha>0, \beta>0 \\ 0, \text{ elsewhere} \end{cases} \tag{4.1.1}$$

A generalized gamma variate is of vital importance in the fields of reliability analysis and life testing models. There is vast literature available on these topics. A number of distributions such as gamma, Weibull, Raleigh, folded normal and negative exponential are special cases of (4.1.1). These distributions are applicable in many problems in physical, social and biological sciences. In most of the cases when (4.1.1) is applied one has to obtain ratios, products and linear combinations of random variables which may be generalized gamma variables or special cases of these variables. Hence in this section we will look into the general problem of products and ratios of generalized gamma variates.

Let $X_1, \ldots, X_m, X_{m+1}, \ldots, X_k$ be a simple random sample of size k from (4 1.1). That is, $X_1, \ldots, X_k$ are random variables independently and identically distributed according to the probability law (4.1.1). Let

$$Y = \frac{X_1 \ldots X_m}{X_{m+1} \ldots X_k}. \tag{4.1.2}$$

If the density $g(y)$ of Y exists, then the $(s-1)$th moment of Y about the origin is given by

$$E(Y^{s-1}) = \prod_{j-1}^{m} \left\{ E\left(X_j^{s-1} \right) \right\} \prod_{j=m+1}^{k} \left\{ E\left(X_j^{-(s-1)} \right) \right\} \tag{4.1.3}$$

$$= \frac{a^{(m-n)/\beta}\, \Gamma^m \left(\frac{\alpha-1}{\beta} + \frac{s}{\beta} \right) \Gamma^{k-m} \left(\frac{\alpha+1}{\beta} - \frac{s}{\beta} \right)}{\Gamma^k \left(\frac{\alpha}{\beta} \right) a^{\frac{(m-n)\,s}{\beta}}}$$

where $n = k - m$ and E denotes *mathematical expectation.* Hence the density function $g(y)$ is available from the inverse Mellin transform of (4.1.3). That is,

$$g(y) = \frac{1}{2\pi i} \frac{a^{(m-n)\cdot\beta}}{\Gamma^k \left(\frac{\alpha}{\beta} \right)} \int_{c-i\infty}^{c+i\infty} \left\{ \Gamma^m \left(\frac{\alpha-1}{\beta} + \frac{s}{\beta} \right) \Gamma^n \left(\frac{\alpha+1}{\beta} - \frac{s}{\beta} \right) a^{(m-n)s/\beta} y^{-s} \right\} ds$$

$$= \frac{a^{(m-n)/\beta}}{\Gamma^k \left(\frac{\alpha}{\beta} \right)} H_{n,m}^{m,n} \left[a^{(m-n)/\beta} y \,\middle|\, \begin{matrix} \left(1-\frac{\alpha+1}{\beta}, \frac{1}{\beta}\right), \ldots, \left(1-\frac{\alpha+1}{\beta}, \frac{1}{\beta}\right) \\ \left(\frac{\alpha-1}{\beta}, \frac{1}{\beta}\right), \ldots, \left(\frac{\alpha-1}{\beta}, \frac{1}{\beta}\right) \end{matrix} \right] \tag{4.1.4}$$

where $i = (-1)^{1/2}$, c is chosen suitably and $0 < y < \infty$, The H-function in (4.1.4) exists when

$$\frac{\alpha-1}{\beta} + \nu \neq -\left(\frac{\alpha+1}{\beta} \right) - r, \ \nu, r = 0, 1, \ldots .$$

By making a transformation of the variable in (4.1.4) we can transform the H-function into a Meijer's G-function.

Some generalizations of (4.1.4) are given in the exercises at the end of this chapter. A general discussion of problems of this type may be seen from Mathai (1972). When (4.1.4) is applied to a practical problem one needs a computable representation in order to compute percentage points. A computable representation of a general H-function was discussed in Chapter 3.

REMARK 1. Instead of the random variables $X_1,\ldots, X_k$ even if we had considered arbitrary powers of these random variables the procedure would have remained the same but, of course, the existence conditions would have changed.

REMARK 2. When dealing with statistical problems it is sufficient to assume the existence to start with and complete the derivation and then impose the existence conditions on the final H-function because the density in these problems is unique whenever it exists.

4.2 LINEAR COMBINATIONS OF RANDOM VARIABLES

In a large variety of statistical problems the distributions of linear combinations of random variables play important roles. For example, consider the total service time required in a medical checkup. The medical examination may consist of several stages such as the blood test, urine test, physical examination, etc. The service time at each stage may be distributed according to some specific distributions such as negative exponential or gamma. Thus the total service time is a linear combination of random variables. In order to make inferential statements about the total service time one needs the distribution of a linear combination of random variables. Such problems abound in the fields of queueing theory and operations research.

Consider the problem of inter-live-birth interval. This is the interval between two live births for a woman in fecundable age groups. In between two live births there may be a number of still births, miscarriages, abortions and associated with these events there are sterile periods. Thus the inter-live-birth interval is a linear combination of several component variables each of which may have its own distribution. For problems of this nature see George and Mathai (1975).

There is a vast literature on the distributions of linear combinations of random variables where the individual variables are assumed to have particular types of distributions. In order to unify as well as to generalize these results Mathai and Saxena (1973) considered the linear combinations of random variables where each component variable is assumed to have a density associated with an H-function so that this density function will contain a large variety of density functions as special cases. Here we list one result for the sake of illustration.

Let $X_1,\ldots, X_n$ be n independent random variables where X_i has the density function

$$f_i(x_i) = \frac{x_i^{\gamma_i - 1}}{C_i} e^{-ap_i x_i} H_{r_i, s_i}^{k_i, l_i}\left[z_i x_i^{\mu_i} \middle| \begin{matrix} (a_{r_i}^{(i)}, A_{r_i}^{(i)}) \\ (b_{s_i}^{(i)}, B_{s_i}^{(i)}) \end{matrix} \right] \tag{4.2.1}$$

for $0 < x_i < \infty$ and $f_i(x_i) = 0$ elsewhere, $i = 1, \ldots, n$; where

$$C_i = (ap_i)^{-\gamma_i} H_{r_i+1, s_i}^{k_i, l_i+1}\left[\frac{z_i}{(ap_i)^{\mu_i}} \middle| \begin{matrix} (-\gamma_i + 1, \mu_i), (a_{r_i}^{(i)}, A_{r_i}^{(i)}) \\ (b_{r_i}^{(i)}, B_{s_i}^{(i)}) \end{matrix} \right] \tag{4.2.2}$$

$$\mu_i > 0,\ R(p_i) > 0,\ R(\gamma_i + 1 + \mu_i \min b_j / B_j) > 0,\ j = 1, \ldots, k_i,$$

$i = 1, 2, \ldots, n$. Here $R(\cdot)$ means the real part of $(\cdot)$. Since there exists at least one set of parameters for which $f_i(x_i)$ in (4.2.1) is non-negative it is assumed that the parameters are such that $f_i(x_i) \geqslant 0$ for $0 < x_i < \infty$.

Let,

$$U = p_1 X_1 + p_2 X_2 + \ldots + p_n X_n \tag{4.2.3}$$

then the density $h(u)$ of U is given as follows.

$$h(u) = \left\{ \prod_{j=1}^{n} \frac{p_i^{-\gamma_i}}{C_i} \right\} \prod_{i=1}^{n} \sum_{h=1}^{k_i} \sum_{\nu_i=0}^{\infty} \prod_{j=1}^{k_i} \Gamma\left[b_j^{(i)} - B_h^{(i)} \frac{(b_h^{(i)} + \nu_i)}{B_j^{(i)}} \right]$$

$$\times \frac{\prod_{j=1}^{l_i} \Gamma\left[1 - a_j^{(i)} + A_j^{(i)} \frac{(b_h^{(i)} + \nu_i)}{B_h^{(i)}} \right] \Gamma\left[\gamma_i + \mu_i \frac{(b_h^{(i)} + \nu_i)}{B_h^{(i)}} \right] \left(\frac{z_i}{p_i^{\mu_i}} \right)^{\frac{b_h^{(i)} + \nu_i}{B_h^{(i)}}}}{\prod_{j=k_i+1}^{s_i} \Gamma\left[1 - b_j^{(i)} + B_j^{(i)} \frac{(b_h^{(i)} + \nu_i)}{B_h^{(i)}} \right] \prod_{j=l_i+1}^{r_i} \Gamma\left[a_j^{(i)} - A_j^{(i)} \frac{(b_h^{(i)} + \nu_i)}{B_h^{(i)}} \right]}$$

$$\times \left\{ \frac{(-1)^{\nu_i}}{\nu_i!\, B_h^{(i)}} \right\} \left[\frac{e^{-au}\, u^{\sum_{j=1}^{n} \left[\mu_j \frac{(b_h^{(j)} + \nu_j)}{B_h^{(j)}} + \nu_j \right] - 1}}{\Gamma\left[\sum_{j=1}^{n} \left\{ \mu_j \frac{(b_h^{(j)} + \nu_j)}{B_h^{(j)}} + \gamma_j \right\} \right]} \right] \tag{4.2.4}$$

The density function in (4.2.4) is derived by finding the moment generating function with the help of the moment generating functions of $X_j, j = 1, \ldots, n$, then collecting appropriate terms and then inverting the moment generating function to get the density.

REMARK 1. In presenting (4.2.4) we have not stated the existence conditions explicitly. These can be worked out from (4.2.1) for the specific parameters under consideration when dealing with a specific problem, see, for example, some problems in the exercises at the end of this chapter.

REMARK 2. In (4.2.1) and (4.2.3) we have taken the same p_i's for simplicity. Also in (4.2.1) the parameter a is assumed to be the same for all j.

These conditions can be deleted. Still the procedure will remain the same but the corresponding expression for $h(u)$ will be more complicated.

4.3 LIKELIHOOD RATIO CRITERIA IN MULTIVARIATE STATISTICAL ANALYSIS

In multivariate statistical analyses often statisticians face the problems of testing various hypotheses on the parameters of a multinormal distribution. There are several standard procedures available for testing such hypotheses. One such procedure is the likelihood ratio principle. A test statistic is constructed with the help of this principle and then the success of the attempt to test the particular hypothesis will depend upon the availability of the exact distribution of the test statistic thereby constructing the exact percentage points. In many such standard problems the density functions of the test statistics can be expressed in terms of H-functions. A few such cases will be mentioned in this section.

Consider the problems of testing the hypothesis that the covariance matrix in a multinormal case is diagonal and the conditional hypothesis that the diagonal elements are equal given that the covariance matrix is diagonal. Let λ be the likelihood ratio criterion for testing the conditional hypothesis and let $W = \lambda^{2/N}$ where N is the sample size. Then the hth moment of W can be shown to be,

$$E(W^h) = \frac{p^{ph}\,\Gamma^p\left(\frac{n}{2}+h\right)\Gamma\left(\frac{np}{2}\right)}{\Gamma^p\left(\frac{n}{2}\right)\Gamma\left(\frac{np}{2}+ph\right)}, \; n = N-1. \tag{4.3.1}$$

Then from the inverse Mellin transform, the density is given by

$$f(w) = \frac{\Gamma\left(\frac{np}{2}\right)}{\Gamma^p\left(\frac{n}{2}\right)} w^{-1} H_{1,p}^{p,0}\left[\frac{w}{p^p}\middle|\begin{matrix}\left(\frac{np}{2}, p\right)\\ \left(\frac{n}{2}, 1\right), \ldots, \left(\frac{n}{2}, 1\right)\end{matrix}\right] \tag{4.3.2}$$

$$0 < w < 1.$$

A one-to-one function of the likelihood ratio criterion for testing sphericity, that is, $\Sigma = \sigma^2 I$ where Σ is the covariance matrix in a multinormal case, I is an identity matrix and σ^2 is an unknown scalar quantity, has the hth moment,

$$E\left(w_1^h\right) = \frac{p^{ph}\,\Gamma\left(\frac{np}{2}\right)}{\Gamma\left(\frac{np}{2}+ph\right)} \prod_{j=1}^{p} \frac{\Gamma\left(\frac{n+1-j}{2}+h\right)}{\Gamma\left(\frac{n+1-j}{2}\right)} \tag{4.3.3}$$

and hence the density function of W_1 is given by

$$f_1(w_1) = \frac{\Gamma\left(\frac{np}{2}\right)}{p^p \prod_{j=1}^{p} \Gamma\left(\frac{n+1-j}{2}\right)} w_1^{-1} \, H_{1,p}^{p,0}\left[\frac{w_1}{p^p}\middle|\begin{matrix}\left(\frac{np}{2}, p\right) \\ \left(\frac{n-1-j}{2}, 1\right) j=1, \ldots, p\end{matrix}\right]$$

$$0 < w_1 < 1. \tag{4.3.4}$$

Consider the problem of testing the hypothesis that the covariance matrices are equal in a set of q independent p-variate multinormal populations. The hth moment of Bartlett's criterion for testing this hypothesis is given by,

$$E(V^h) = \prod_{i=1}^{p} \prod_{g=1}^{q} \frac{\Gamma\left(\frac{n_g + hn_g + 1 - i}{2}\right)}{\Gamma\left(\frac{n_g + 1 - i}{2}\right)} \frac{\Gamma\left(\frac{n+1-i}{2}\right)}{\Gamma\left(\frac{n + hn + 1 - i}{2}\right)} \tag{4.3 5}$$

Hence the density of V is given by

$$g(v) = \prod_{i=1}^{p} \prod_{g=1}^{q} \frac{\Gamma\left(\frac{n+1-i}{2}\right)}{\Gamma\left(\frac{n_g+1-i}{2}\right)} v^{-1} \tag{4.3.6}$$

$$H_{p,pq}^{pq,0}\left[v\middle|\begin{matrix}\left(\frac{n+1-i}{2}, \frac{n}{2}\right) & i=1, \ldots, p \\ \left(\frac{n_g+1-i}{2}, \frac{n_g}{2}\right), & j=1, \ldots, q, i=1, \ldots, p\end{matrix}\right], 0 < u < 1$$

where $n_g = N_g - 1$, $n = \sum_{g=1}^{q} n_g$ and N_g, $g = 1, \ldots, q$ are the sample sizes.

There are other similar problems connected with the likelihood ratio criteria where the densities can be put in terms of H-functions.

REMARK. In all the above problems the densities of the statistics or one-to-one functions of these statistics can be reduced to Meijer's G-functions by the application of Gauss-Legendre multiplication formula for the gamma function.

4.4 GENERALIZED PROBABILITY LAWS

A class of probability models may have some common properties. It is often convenient to study such properties for the general class rather than repeating the study for each and every member of the class. For example, consider the generalized gamma density given in (4.1.1). Here the probability models—gamma, exponential and folded normal—are all special cases. Hence by studying the properties of (4.1.1) we obtain some of the properties enjoyed by the individual members of this class. There may be some properties which are enjoyed by a special member and not enjoyed by the general class.

Generalized classes of probability models were considered by several authors from time to time. One such example is the Pearson's system of probability models which are generated from a differential equation involving an arbitrary density function. Mathai and Saxena (1966) introduced a general class of probability models which were defined in terms of a hypergeometric function ${}_2F_1(\cdot)$. Some generalizations of their results to matrix variate cases are considered by Roux (1975). Some of these results will be discussed in Chapter 5.

Since the H-function being a very general special function, it can define a very general class of probability models. The parameters are to be restricted in such a way that the function is nonnegative and finite in the region under consideration. One such general probability model is defined by Mathai and Saxena (1971b) as follows:

$$f(x) = \frac{e^{-dx}\,x^{\lambda-1}}{C\,(1+bx^k)^{\mu-1}}\; H^{m,n}_{p,q}\left[\left(\frac{ax^k}{1+bx^k}\right)^s \middle| \begin{matrix}(a_1, A_1), \ldots, (a_p, A_p)\\ (b_1, B_1), \ldots, (b_q, B_q)\end{matrix}\right], \tag{4.4.1}$$

for $x > 0$ and $f(x) = 0$ elsewhere, where

$$C = \sum_{r=0}^{\infty} \frac{(-1)^r\,d^r}{r!\;k}\, b^{-(\lambda+r)/k}\, \Gamma\left(\mu - 1 - \frac{\lambda+r}{k}\right)$$

$$\times H^{m,n+1}_{p+1,q+1}\left[\left(\frac{a}{b}\right)^s \middle| \begin{matrix}\left(1-\left(\frac{\lambda+r}{k}\right), s\right), (a_1, A_1), \ldots, (a_p, A_p)\\ (b_1, B_1), \ldots, (b_q, B_q), (2-\mu, s)\end{matrix}\right] \tag{4.4.2}$$

Obviously $f(x)$ is not a nonnegative function for all possible values of the parameters. But there exists a number of sets of parameters for which

$f(x) > 0,\ 0 < x < \infty$ and $\int_0^{\infty} f(x)\,dx = 1$. Hence $f(x)$ in (4.4.1) is restricted to those parameter values. It can be shown that a number of central and noncentral distributions, which are frequently used in statistical analysis, are special cases of (4.4.1). Some of them are listed in the exercises at the end of this chapter. Some properties of (4.4.1) are studied in Mathai and Saxena (1971b).

REMARK. When (4.4.1) is applied to practical problems one may have to restrict to specific parameter values. But in general (4.4.1) will help one to look for more general cases which can describe a practical situation which may be more general than the one under consideration of the experimenter.

4.5 CHARACTERIZATIONS OF PROBABILITY LAWS

The topic of characterization of probability laws is a vast and growing field of mathematical statistics and probability theory. There are essentially two main areas in this field: One is the investigation of properties by which a probability law can be uniquely determined. For example, it can be shown that under certain minor conditions that the only distribution where the sample mean and the sample variance are independently distributed is the normal distribution. In other words, under certain conditions, the independence of the sample mean and the sample variance characterizes the normal distribution.

Another type of characterization is the investigation of some postulates by which one can give an axiomatic definition to some basic concepts in statistics. This is the topic of characterization of fundamental concepts. Recent developments in this field are discussed in Mathai and Rathie (1975).

Since a *H*-function is a very general function it is obvious that the properties of this function cannot be used for the unique determination of any specific probability law. But a general function such as a *H*-function can be profitably used in solving problems of this nature from a different angle.

A particular property may be suspected of being the unique property of a particular random variable. But a general function may be used to show that the property is enjoyed by a general class of probability laws thus disproving the claim that the property is uniquely enjoyed by a particular distribution. For example, consider the following ratio property. It is well known that the ratio of two independent standard normal variables has a Cauchy distribution. Now one may ask the question that if the ratio of two independent random variables is known to be a Cauchy random variable are those two variables necessarily be normal variables? Questions of this nature have been answered by many workers and a general technique of tackling such problems is given in Mathai and Saxena (1969a). The essential idea in this paper is the following: First a general class of probability laws associated with a *H*-function is studied. Then a connection between this general class and a particular ratio distribution, such as Cauchy or Student-t or variance ratio, is established with the help of the Parseval property of Hankel transform. Thus essentially showing that if any two independent random variables, having probability laws in the general class, are taken then their ratio will have the particular ratio distribution under consideration. This obviously disproves the conjecture that the variables are uniquely determined by the particular ratio property. For details of problems of this type see Mathai and Saxena (1969a).

4.6 STRUCTURAL SETUP OF PROBABILITY LAWS

Another area in statistics where the properties of H-function can be applied is the field of examination of general structures of statistics (certain functions of random variables). In practical applications one often needs the distributions of certain functions of random variables. For example, in the problems of tests of statistical hypotheses when test statistics are constructed, these statistics are functions of the sample. These statistics may be complicated functions of the sample but structurally they may be considered as products or ratios of independent random variables which are known. In that case one may be able to get the $(s-1)$th moment of the statistic, which is nothing but the Mellin transform of the density of the statistic, without knowing the density itself. By making a comparison of this moment with the Mellin transform of a G or H-function one may be able to obtain the density function as a G or H-function.

Structural analysis of a test statistic is often a powerful tool in deriving its density. Some problems of this nature are tackled in Mathai (1972) and Mathai and Saxena (1969). For the sake of illustration we will examine one test statistic coming from a statistical test in multivariate statistical analysis.

A one-to-one function of Wilks' criterion for testing regression has the $(s-t)$th moment,

$$E(U^{s-1}) = \prod_{j=1}^{p} \frac{\Gamma\left(\frac{n+1+q-j}{2}\right)}{\Gamma\left(\frac{n+1-j}{2}\right)} \prod_{j=1}^{p} \frac{\Gamma\left(\frac{n+1-j}{2}+s-1\right)}{\Gamma\left(\frac{n+1+q-j}{2}+s-1\right)} \tag{4.6.1}$$

By examining (4.6.1) one can easily see that if the $(s-1)$th moment of a product of p independent beta random variables is taken then one can get the moment having the same structure as in (4.6.1). Thus one can consider U, structurally, as the product of p independent beta random variables. In order to find the density of U one needs to find the density of the product of independent beta random variables having certain parameters. The exact density of the product of independent beta variables with arbitrary parameters is available from Mathai (1971b).

By examining (4.6.1) one can also see that the right side of (4.6.1) is nothing but the Mellin transform of a G-function which is a special case of a H-function. Thus the density of U is available in terms of a Meijer's G-function or a particular case of a H-function. Problems of this type are discussed in Mathai and Saxena (1973a, Chapters 5 and 6).

4.7 OTHER APPLICATIONS

It goes without saying that most of the practical problems, where elementary special functions are used, are capable of being generalized with the help of generalized special functions such as a *H*-function. Such generalizations may be mathematically interesting but they often do not lead to meaningful physical interpretations. Problems of heat conduction and limiting of sinusoidal signals, which are discussed in Mathai and Saxena (1973a, Chapter 7), can be given such generalizations by altering some basic conditions.

McNolty and Tomsky (1972) consider the problem of phase and radial densities Phase distributions arise in the problem of reconstructing a phase density function from a given radial or amplitude density. Saxena and Sethi (1973a) have generalized these problems with the help of distributions associated with generalized hypergeometric functions. Further generalizations, with the help of *H*-functions, is possible. As long as one can find practical situations, where the introduction of a more general function is justifiable, these generalizations can be put to practical use.

Queueing theory is another area where one can apply various special functions. Even with very elementary distributions, such as a Poisson distribution for the arrivals, the problem becomes complicated depending upon the nature of the process itself. Even if the service time components are assumed to be independent negative exponentials with different expected durations the total service time distribution itself is complicated. Sometimes it is meaningful to assume a gamma or a generalized gamma distribution for the service time components. If the total service time is a simple sum of independent such components then the total service time will have the structure

$$T = X_1 + \ldots + X_k \tag{4.7.1}$$

where $X_1, \ldots, X_k$ are independent generalized gamma variables. We will need the properties of *H*-functions in order to evaluate the density of T. Some general problems of this nature are discussed in Mathai and Saxena (1973).

In the above situation if we would like to interpret and estimate coefficients such as "traffic intensity" the resulting estimators will have complicated distributions even if we use estimators which are derived with the help of some standard techniques such as the method of moments. These are problems where *H*-functions come in as a natural aftermath of the nature of the problem itself resulting from some justifiable modifications of the basic assumptions involved in the problem.

EXERCISES

4.1 If $X_1, \ldots, X_k$ are independent random variables having the density functions

$$f_j(x_j) = \frac{\beta_j a_j^{\alpha_j/\beta_j}}{\Gamma\left(\frac{\alpha_j}{\beta_j}\right)} x_j^{\alpha_j - 1} e^{-a_j x_j^{\beta_j}},\ x_j > 0,\ a_j > 0,\ \alpha_j > 0,\ \beta_j > 0,\ j = 1, 2, \ldots, k,$$

and if $Y = \frac{X_1 \ldots X_m}{X_{m+1} \ldots X_k}$ then show that the density function $g(y)$ of Y, whenever it exists, is given by

$$g(y) = C\, y^{-1} H^{m,n}_{n,m}\left[\left(\frac{a}{a'}\right) y \,\middle|\, \begin{matrix} \left(1 - \frac{\alpha_{m+1}}{\beta_{m+1}}, \frac{1}{\beta_{m+1}}\right), \ldots, \left(1 - \frac{\alpha_k}{\beta_k}, \frac{1}{\beta_k}\right) \\ \left(\frac{\alpha_1}{\beta_1}, \frac{1}{\beta_1}\right), \ldots, \left(\frac{\alpha_m}{\beta_m}, \frac{1}{\beta_m}\right) \end{matrix}\right]$$

where

$$C = \left\{\prod_{j=1}^{k} \Gamma\left(\frac{\alpha_j}{\beta_j}\right)\right\}^{-1},\ n = k - m,\ a = \prod_{j=1}^{m} a_j^{1/\beta_j},$$

$$a' = \prod_{j=m+1}^{k} a_j^{1/\beta_j},\ 0 < y < \infty.$$

4.2 Show that the density function in 4.1 above can be written in terms of a Meijer's G-function in the following cases: (1) $\beta_1 = \beta_2 = \ldots = \beta_k = \beta$; (2) $\beta_1, \ldots, \beta_k$ rational.

4.3 In (4.1.4), when $m = 2$, $n = 0$, show that

$$g(y) = C\, z^{(\alpha-1)/\beta} K_0(2 z^{1/2})$$

where C is a normalizing constant, $z = a^{m-n} y^{\beta}$ and $K_n(z)$ is a modified Bessel function of integer order. (Mathai, 1972)

4.4 In (4.1.4), when $m = 2$, $n = 1$, show that the density function $g(y)$ is given by

$$g(y) = C\, z^{-(\alpha-1)/\beta} \sum_{\nu=0}^{\infty} \frac{z^{\nu}\, \Gamma\left(\frac{2\alpha}{\beta} + \nu\right)}{(\nu!)^2} \left[-\log z + 2\psi(\nu + 1) + \psi\left(\frac{2\alpha}{\beta} + \nu\right)\right]$$

where C is a normalizing constant, $z = a^{m-n} y^{\beta}$ and $\psi(\cdot)$ is a ψ-function. (Mathai, 1972)

4.5 If $X_1, \ldots, X_n$ are independent random variables having the density functions

$$J_i(x_i) = \left(\frac{1}{E_i}\right) e^{-ap_i x_i}\, x_i^{\gamma_i - 1}\; {}_2F_1(\alpha_i, \beta_i; \delta_i; -z_i x_i^{\mu_i}),$$

$0 < x_i < \infty$ and zero elsewhere, where

$$E_i = (aP_i)^{-\gamma_i}\, H^{1,3}_{3,2}\left[\frac{z_i}{(p_i a)^{\mu_i}}\middle| \begin{matrix}(\gamma_i + 1, \mu_i), (1-\alpha_i, 1), (1-\beta_i, 1)\\ (0, 1), (1-\delta_i, 1)\end{matrix}\right]$$

show that the density function of $U = p_1X_1 + \ldots + P_nX_n$ is given by

$$h(u) = \prod_{i=1}^{n}\left\{P_i^{-\gamma_i}/E_i\right\} \prod_{i=1}^{n}\left[\sum_{\nu_i=0}^{\infty} \frac{\Gamma(\alpha_i+\nu_i)\,\Gamma(\beta_i+\nu_i)\,\Gamma(\gamma_i+\mu_i\nu_i)}{\Gamma(\delta_i+\nu_i)}\right.$$

$$\left.\times \frac{(-1)^{\nu_i}}{\nu_i!}\right] \frac{e^{-au}\, u^{\sum_{j=1}^{n}(\mu_j\nu_j+\gamma_j)-1}}{\Gamma\left[\sum_{j=1}^{n}\{\mu_j\nu_j+\nu_j+\gamma_j\}\right]}.$$

HINT. Put $k_i = 1,\ l_i = 2,\ r_i = s_i = 2,\ A_i^{(i)} = 1 = A_2^{(i)} = B_1^{(i)} = B_2^{(i)}$,

$a_1^{(i)} = 1-\alpha_i,\ a_2^{(i)} = 1-\beta_i,\ b_1^{(i)} = 0,\ b_2^{(i)} = 1-\delta_i$ for $i = 1,\ldots,n$ in (4.2.4).

(Mathai and Saxena, 1973)

4.6 If $X_1, \ldots, X_n$ are independent random variables with density functions

$$q_j(x_j) = \left(\frac{1}{F_j}\right) e^{-ap_j x_j}\, x_j^{\gamma_j-1}\; {}_1F_1(\alpha_j; \delta_j; -z_j x_j^{\mu_j}),$$

$$0 < x_j < \infty,\ j = 1, \ldots, n,\ a, p_j, \gamma_j, \alpha_j, \delta_j, z_j, \mu_j > 0$$

where

$$F_j = (ap_j)^{-\gamma_j}\, H^{1,2}_{2,2}\left[\frac{z_j}{(ap_j)^{\mu_j}}\middle| \begin{matrix}(-\gamma_j+1, \mu_j), (1-\alpha_j, 1)\\ (0, 1), (1-\delta_j, 1)\end{matrix}\right]$$

and if $U = p_1X_1 + \ldots + p_nX_n$ then show that the density function of U is given by,

$$h(u) = \prod_{i=1}^{n}\left\{\frac{p_i^{-\gamma_i}}{F_i}\right\} \prod_{i=1}^{n}\left[\sum_{\nu_i=0}^{\infty} \frac{\Gamma(\alpha_i+\nu_i)\,\Gamma(\gamma_i+\mu_i\nu_i)\,(-1)^{\nu_i}}{\Gamma(\delta_i+\nu_i)\quad \nu_i!}\right.$$

$$\left.\times \frac{e^{-au}\, u^{\sum_{j=1}^{n}(\mu_j\nu_j+\gamma_j)-1}}{\Gamma\left[\sum_{j=1}^{n}(\mu_j\nu_j+\gamma_j+\nu_j)\right]}\right]$$

(Mathai and Saxena, 1973)

4.7 Prove the following results.

(a) $$2^{n-1}\, H_{1,2}^{2,0}\left[\frac{z}{4}\middle|\begin{matrix}(n, 2)\\ \left(\frac{n}{2}, 1\right), \left(\frac{n}{2}, 1\right)\end{matrix}\right] = \pi^{1/2}\, G_{1,1}^{1,0}\left[z\middle|\begin{matrix}\frac{n}{2}+\frac{1}{2}\\ \frac{n}{2}\end{matrix}\right]$$
$$= z^{n/2}(1-z)^{-1/2},\ 0<|z|<1.$$

(b) $$3^{(3n-1)/2}(2\pi)^{-1} H_{1,3}^{3,0}\left[\frac{z}{9}\middle|\begin{matrix}\left(\frac{3n}{2}, 3\right)\\ \left(\frac{n}{2}, 1\right), \left(\frac{n}{2}, 1\right), \left(\frac{n}{2}, 1\right)\end{matrix}\right]$$
$$= G_{2,2}^{2,0}\left[z\middle|\begin{matrix}\frac{n}{2}+\frac{1}{3}, \frac{n}{2}+\frac{2}{3}\\ \frac{n}{2}, \frac{n}{2}\end{matrix}\right] = z^{n/2}\, {}_2F_1\left(\frac{2}{3}, \frac{1}{3}; 1; 1-z\right),$$
$$0<|z|<1.$$

(c) $$2^{4n-5/2}\,\pi^{-3/2}\, H_{1,4}^{4,0}\left[\frac{z}{16}\middle|\begin{matrix}(2n, 4)\\ \left(\frac{n}{2}, 1\right), \left(\frac{n}{2}, 1\right), \left(\frac{n}{2}, 1\right), \left(\frac{n}{2}, 1\right)\end{matrix}\right]$$
$$= G_{3,3}^{3,0}\left[z\middle|\begin{matrix}\frac{n}{2}+\frac{1}{4}, \frac{n}{2}+\frac{2}{4}, \frac{n}{2}+\frac{3}{4}\\ \frac{n}{2}, \frac{n}{2}, \frac{n}{2}\end{matrix}\right],\ 0<|z|<1.$$

(Mathai, 1972)

4.8 Prove the following result.

$$y^{-1} H_{1,3}^{3,0}\left[\frac{y}{9}\middle|\begin{matrix}(3a, 3)\\ a, a-1/2, a-1\end{matrix}\right]$$
$$= 2\pi\, 3^{-3a+1/2}\, G_{2,2}^{2,0}\left[y\middle|\begin{matrix}a-2/3, a-1/3\\ a-3/2, a-2\end{matrix}\right]$$
$$= 2^3\,\pi^{1/2}\, 3^{-3a-1/2}\, y^{a-2}(1-y)^{3/2}\, {}_2F_1\left(\frac{5}{6}, \frac{7}{6}; \frac{5}{2}; 1-y\right),$$
$$0<|y|<1.$$

4.9 Show that the noncentral chi-square distribution with the density function,

$$f_1(x) = e^{-\mu^2/2\sigma^2}\sum_{r=0}^{\infty}\frac{1}{r!\,\Gamma\left(r+\frac{k}{2}\right)}\left(\frac{\mu^2}{2\sigma^2}\right)^r\left(\frac{1}{2}\right)^{r+k/2} e^{-x/2}\, x^{r+k-1}$$

for $x>0$, is available from (4.4.1) as a special case.

4.10 By specializing the parameters in (4.4.1) derive the noncentral F density, namely

$$f_2(x) = e^{-\lambda^2/2} \sum_{r=0}^{\infty} \frac{(\lambda^2)^r}{2} \frac{\Gamma\left(\frac{k+m}{2}+r\right)}{r!\,\Gamma\left(\frac{k}{2}+r\right)\Gamma\left(\frac{m}{2}\right)} \frac{x^{r-1+k/2}}{(1+x)^{r+(k+m)/2}}$$

for $x > 0$.

Hence or otherwise deduce the noncentral beta density from (4.4.1).

4.11 By specializing the parameters in (4.4.1) derive the noncentral Student-t distribution,

$$f_3(x) = \frac{\nu^{\nu/2}\, e^{-\delta^2/2} \sum_{r=0}^{\infty} \Gamma\left(\frac{\nu+1+r}{2}\right)\left(\frac{\delta^r}{r!}\right)\left(\frac{2x^2}{\nu+x^2}\right)^{r/2}}{\Gamma\left(\frac{\nu}{2}\right)(\nu+x^2)^{(\nu+1)/2}}, \quad -\infty < x < \infty$$

in its folded form, that is, $2f_3(x)$ for $x > 0$.

4.12 By specializing the parameters in (4.4.1) derive the density function associated with a hypergeometric function, namely,

$$f_4(x) = \frac{b\, a^{c/b}\, \Gamma(\alpha)\, \Gamma(\beta)\, \Gamma\left(\gamma - \frac{c}{b}\right) x^{c-1}}{\Gamma\left(\frac{c}{b}\right)\Gamma(\gamma)\,\Gamma\left(\alpha - \frac{c}{b}\right)\Gamma\left(\beta - \frac{c}{b}\right)}\, {}_2F_1(\alpha, \beta; \gamma; -ax^b)$$

$$x > 0,\ c > 0,\ \alpha - \frac{c}{b} > 0,\ \beta - \frac{c}{b} > 0.$$

CHAPTER 5

Special Functions of Matrix Argument

In this chapter we will define the various special functions when the argument is a symmetric positive definite matrix. After defining the elementary functions such as exponential, gamma, beta, etc. the definition is extended to cover Meijer's *G*-function. A definition for the *H*-function of matrix argument is also given with the help of a generalized matrix transform.

Integral representations of elementary special functions are mainly available through generalized Laplace transforms and series representations are available through zonal polynomials. Some notations and preliminary results are given in Section 5.1. The various functions, some of their properties, some integrals involving these functions and some applications to statistical problems are given in the remaining sections of this chapter.

The developments of the definitions and properties are mainly based on the papers of Herz (1955) for elementary functions through integral transforms, James (1964) for series representations through zonal polynomials, Mathai and Saxena (1971a) and Mathai (1976) for *G* and *H*-functions of matrix argument.

As far as possible this chapter is made self-contained but some basic knowledge of linear Algebra is essential and the other topics may be picked up without much difficulty. The development of the subject matter will be mainly through Laplace and generalized Mellin transforms. This approach has the defect that explicit expressions for the various special functions are not available even though their properties can be studied easily. The alternate approach, namely, the approach through zonal polynomials, enables one to obtain explicit expressions in some special cases but a number of properties cannot be established through this approach due to the obvious fact that a series representation has limitations when it comes to integration, differentiation, etc.

5.1 SOME NOTATIONS AND PRELIMINARY RESULTS

The notations which are going to appear frequently are listed here for convenience. Other notations are given wherever they occur for the first time. For square matrices P and Q

$P > 0$ means that P is positive definite; (5.1.1)
$P > Q$ means that $P - Q$ is positive definite;
$P \geqslant 0$ means that P is positive semi-definite;
$\det P \equiv$ determinant of $P \equiv |P|$;

$\int_{P>0} (\ )$ means that the integral is over the set of positive definite matrices P;

$\int_{0(m)} dH$ means that the integral is over the orthogonal group with respect to the Haar measure dH.

$$\Gamma_m(\delta) = \pi^{m(m-1)/4}\,\Gamma(\delta)\,\Gamma(\delta - \tfrac{1}{2}) \ldots \Gamma\left(\delta - \frac{m-1}{2}\right) \tag{5.1.2}$$

$$\Pi_m(\delta) = \Gamma_m\left(\delta + \frac{m+1}{2}\right)$$

$$\text{etr}\ Q = e^{tr\ Q} \tag{5.1.3}$$

where $tr\ Q$ = trace of Q = sum of the leading diagonal elements of Q. That is, if $Q = (q_{ij})$ is a square matrix of order k where q_{ij} is the ith row jth column element then $tr\ Q = q_{11} + q_{22} + \ldots + q_{kk}$.

5.1.1 Some Results on Jacobians of Matrix Transformations

In the discussion of the various results on integrals involving matrix arguments often we will have to employ transformations of variables. Hence we will need various results on the relationship between the differential elements. When the argument is a square matrix of order k one can treat the problem as a multivariable case involving k^2 scalar variables or $k(k+1)/2$ scalar variables if the matrix is symmetric.

If $A = (a_{ij})$ is a matrix then for convenience we will denote the differential element $da_{11}\, da_{12} \ldots da_{1k}\, da_{21} \ldots da_{2k} \ldots da_{kk}$ by dA. From the theory of Jacobians of transformations one can derive the following relationships without much difficulty.

$$y = Ax \Rightarrow dy = (\det A)\, dx \tag{5.1.4}$$

where y and x are vectors of order k each and A is a nonsingular matrix.

$$Y = AX \Rightarrow dY = (\det\ A)^q dX \tag{5.1.5}$$

where X and Y are $p \times q$ matrices and A is a $p \times p$ nonsingular matrix.

$$Y = AXB \Rightarrow dY = (\det A)^q (\det B)^p \, dX \tag{5.1.6}$$

where X and Y are $p \times q$ matrices, A is a $p \times p$ and B is a $q \times q$ nonsingular matrices.

$$Q_2 = TQ_1T' \Rightarrow dQ_2 = (\det T)^{p+1} \, dQ_1 \tag{5.1.7}$$

where Q_1, Q_2 are $p \times p$ symmetric matrices and T is a $p \times p$ nonsingular matrix.

$$V = TT' \Rightarrow dV = 2^p \prod_{i=i}^{p} t_{ii}^{p+1-i} \tag{5.1.8}$$

where V is a $p \times p$ symmetric matrix and $T = (t_{ij})$ is a triangular matrix.

$$L = AMA' \Rightarrow dL = \prod_{i=1}^{p} a_{ii}^{i} \, dM \tag{5.1.9}$$

where L, M and A are $p \times p$ lower triangular matrices.

$$A = B^{-1} \Rightarrow dA = (\det B)^{-2p} \, dB \tag{5.1.10}$$

where A and B are $p \times p$ nonsingular matrices. If A is a symmetric matrix then $dA = (\det B)^{-(p+1)} \, dB$.

A few more results on Jacobians are given after Definition 5.1.5.

DEFINITION 5.1.1 The differential operators D_Λ and D_Z are defined as,

$$D_\Lambda = \det\left(\eta_{ij} \frac{\partial}{\partial \lambda_{ij}}\right) \quad \text{and} \quad D_Z = \det\left(\frac{\partial}{\partial z_{ij}}\right) \tag{5.1.11}$$

for $\Lambda \in S_m$ and $Z \in S_m^*$ respectively, where S_m is the space of all real $m \times m$ symmetric matrices $\Lambda = (\lambda_{ij})$, $i, j = 1, 2, \ldots, m$ and S_m^* is the corresponding space of $Z = (\eta_{ij} z_{ij})$, $\eta_{ij} = 1$ if $i = j$ and $\frac{1}{2}$ otherwise. The following property is evident from the definition itself.

$$D_\Lambda \operatorname{etr}(\Lambda Z) = (\det Z) \operatorname{etr}(\Lambda Z) \tag{5.1.12}$$

$$D_Z \operatorname{etr}(\Lambda Z) = (\det \Lambda) \operatorname{etr}(\Lambda Z)$$

DEFINITION 5.1.2 *Zonal Polynomials.* Let V_k be the vector space of homogeneous polynomials $\phi(\Lambda)$ of degree k in the $n = m(m+1)/2$ different elements of the $m \times m$ symmetric matrix Λ. The dimension N of V_k is the number $N = (n + k - 1)! / [(n-1)! \, k!]$ of monomials $\prod_{i \leqslant j}^{m} \lambda_{ij}^{k_{ij}}$ of degree $\sum_{i \leqslant j}^{m} k_{ij} = k$.

Consider a congruence transformation $\Lambda \to L\Lambda L'$ by a nonsingular $m \times m$ matrix L. Then we can define a linear transformation of the space V_k of polynomials $\phi(\Lambda)$, that is, $\phi \to L\phi : (L\phi)(\Lambda) = \phi(L^{-1}\Lambda L^{-1\prime})$. A subspace $V' \subset V_k$ is called invariant if $LV' \subset V'$ for all nonsingular matrices L. Also V' is called an irreducible invariant subspace if it has no proper invariant subspace. It can be shown that V_k decomposes into a direct sum of irreducible invariant subspaces V_K corresponding to each partition $K = (k_1, k_2, \ldots, k_m)$, $k = k_1 + \ldots + k_m$ into not more than m parts $V_k = \underset{K}{\oplus} V_K$. The polynomial $(\text{tr}\,\Lambda)^k \in V_k$ then has a unique decomposition $(\text{tr}\,\Lambda)^K = \sum_K C_K(\Lambda)$ into polynomials $C_K(\Lambda) \in V_K$ belonging to the respective subspaces. The zonal polynomial $C_K(\Lambda)$ is thus a symmetric homogeneous polynomial of degree k in the latent roots of Λ.

Zonal polynomials have been developed by Hua (1959) and James (1961). A more detailed discussion may be found in these papers. The following properties are immediate consequences of the definition itself. When Λ is a 1×1 matrix, namely, a scalar quantity x, then

$$C_K(x) = x^k \tag{5.1.13}$$

That is, $C_K(\Lambda)$ corresponds to x^k in the one dimensional case.

For the identity matrix I_m it can be shown that,

$$C_K(I_m) = \chi_{[2K]}(1)\frac{2^k\,k!}{(2k)!}\,Z_K(I_m),\; Z_K(I_m) = 2^k\left(\frac{m}{2}\right)_K \tag{5.1.14}$$

where $\chi_{[2K]}(1)$ is the dimension of the representation $[2K]$ of the symmetric group on $2k$ symbols, where, in general $(a)_K$ means the following:

$$(a)_K = \prod_{i=1}^{m} (a - \tfrac{1}{2}(i-1))_{k_i} \tag{5.1.15}$$

and

$$(a)_{k_i} = a(a+1)\ldots(a+k_i-1).$$

If Λ is a symmetric matrix of rank $h < m$ then $C_K(\Lambda) = 0$ if $k_{h+1} \neq 0$. Also $C_K(\Lambda) \equiv 0$ for partitions K into more than m nonzero parts.

$$\int_{0(m)} (\text{tr}(XH))^{2k}(dH) = \sum_K \frac{(\frac{1}{2})_K}{(m/2)_K}\, C_K(XX') \tag{5.1.16}$$

and

$$\int_{0(m)} C_K(\Lambda HTH')(dH) = \frac{C_K(\Lambda)\,C_K(T)}{C_K(I_m)} \tag{5.1.17}$$

where (dH) stands for the invariant or Haar measure on the orthogonal group $0(m)$, normalized so that the measure of the whole group is unity.

This invariant measure on $0(m)$ is proportional to the area of the $m(m-1)/2$ dimensional hypersurface in Euclidean m^2—space defined by the $m(m+1)/2$ equations $HH' = I_m$, in the m^2 components of H and $\int_{0(m)} d(H) = 2^m \pi^{m^2/2}/\Gamma_m(m/2)$.

DEFINITION 5.1.3 *Laplace and Inverse-Laplace Transforms of Matrix Variables.* Consider the equation

$$\int_{\Lambda>0} \text{etr}(-\Lambda Z) f(\Lambda)\, d\Lambda = g(Z) \tag{5.1.18}$$

The L.H.S. of (5.1.18) is the integral with respect to the measure $d\Lambda$ on S_m extended over the set of all positive definite matrices. If it is absolutely convergent in some right-half-plane $R(Z) > X_0$ then it represents the complex analytic function $g(Z)$ known as the Laplace transform of $f(\Lambda)$ where $R(\cdot)$ means the real part of $(\cdot)$ and X_0 is a fixed value of X and $Z = X + iY$, $i = \sqrt{-1}$. If $\int_{S_m^*} |g(X+iY)|\, dY < \infty$ for some $X > X_0$, the same is true for all $X > X_0$ and then,

$$\frac{1}{(2\pi i)^{m(m+1)/2}} \int_{R(Z)=X} \text{etr}(\Lambda Z)\, g(Z)\, dZ = \begin{cases} f(\Lambda), & \Lambda > 0 \\ 0, & \text{elsewhere.} \end{cases} \tag{5.1.19}$$

Here the integration is taken over $Z = X + iY$ with X fixed $> X_0$ and Y over S_m^*. Thus (5.1.19) gives the inverse Laplace transform.

As an illustration of (5.1.18) and (5.1.19) we shall define the gamma and beta functions of matrix variables.

DEFINITION 5.1.4 *The Generalized Gamma Function.*

$$\int_{\Lambda>0} \text{etr}(-\Lambda Z)\, (\det \Lambda)^{\delta-(m+1)/2}\, d\Lambda = \Gamma_m(\delta)\, (\det Z)^{-\delta} \tag{5.1.20}$$

The integral is absolutely convergent for $R(Z) > 0$, $R(\delta) > (m+1)/2 - 1$. For real $Z > 0$ we have the unique positive definite square root of Z and hence by making a transformation to $Z^{-1/2}\Lambda Z^{-1/2}$ and recalling that the differential elements are such that $d(Z^{-1/2}\Lambda Z^{-1/2}) = (\det Z)^{-(m+1)/2}\, d\Lambda$ the L.H.S. of (5.1.20) reduces to

$$\int_{\Lambda>0} \text{etr}(-\Lambda Z)\, (\det \Lambda)^{\delta-(m+1)/2}\, d\Lambda \tag{5.1.21}$$

$$= \int_{\Lambda>0} \operatorname{etr}(-\Lambda)\,(\det \Lambda)^{\delta-(m+1)/2}\,(\det Z)^{-\delta}$$

Since $\operatorname{etr}(-\Lambda Z)\,(\det \Lambda)^{\delta-(m+1)/2}$ is absolutely integrable for $R(Z)>0$ when $R[\delta-(m+1)/2]>-1$ it is square integrable when $R[\delta-(m+1)/2]>-1/2$ or when $R(\delta)>m/2$. Hence $\det(X+iY)^{-\delta}$ is square integrable in Y for fixed $X>0$ when $R(\delta)>m/2$ and hence absolutely integrable for $R(\delta)>m$. Therefore

$$\frac{1}{(2\pi i)^{m(m+1)/2}} \int_{R(Z)=X_0>0} \operatorname{etr}(\Lambda Z)\,(\det Z)^{-\delta-(m+1)/2}\,dZ \tag{5.1.22}$$

$$= \begin{cases} \dfrac{(\det \Lambda)^{\delta}}{\Pi_m(\delta)}, & \Lambda>0 \\ 0, & \text{elsewhere} \end{cases}$$

where

$$\Pi_m(\delta) = \Gamma_m[\delta+(m+1)/2] \quad \text{and} \quad R(\delta)>[(m+1)/2-1].$$

By direct integration of the L.H.S. of (5.1.20) we can show that $\Gamma_m(\delta)$ has the expression given in (5.1.2). For convenience take $Z=I$ and then make a transformation $\Lambda=TT'$ where T is a lower triangular matrix. Thus from (5.1.8) we get

$$d\Lambda = 2^m \prod_{i=1}^{m} t_{ii}^{m+1-i} \tag{5.1.23}$$

and

$$(\det \Lambda) = (\det T)\,(\det T') = t_{11}^2\, t_{22}^2 \ldots t_{mm}^2 .$$

Since $\Lambda>0$ we have $t_{ii}>0$ for $i=1,\ldots,m$ and t_{ij} for $i \neq j$ are such that $-\infty<t_{ij}<\infty$. Also $tr\,\Lambda$ in this case becomes

$$tr\,TT' = t_{11}^2 + (t_{21}^2+t_{22}^2) + \ldots + (t_{m1}^2+t_{m2}^2+\ldots+t_{mm}^2) \tag{5.1.24}$$

Hence

$$\int_{\Lambda>0} (\det \Lambda)^{\delta-(m+1)/2}\, e^{-tr\,\Lambda}\, d\Lambda \tag{5.1.25}$$

$$= 2^m \left\{ \prod_{i=1}^{m} \int_0^\infty t_{ii}^{2[\delta-(m+1)/2]}\, t_{ii}^{m+1-i}\, e^{-t_{ii}^2}\, dt_{ii} \right\} \left\{ \int_0^\infty e^{-t^2}\, dt \right\}^{m(m-1)/2}$$

$$= \pi^{m(m-1)/4} \prod_{i=1}^{m} \Gamma\left(\delta - \frac{i-1}{2}\right). \tag{5.1.26}$$

The convolution property of Laplace transform can also be generalized to the case of matrix argument.

Let g_1 and g_2 be the Laplace transforms of f_1 and f_2 respectively. Then $g_1 g_2$ is the Laplace transform of f_3, where

$$f_3(\Lambda) = \int_0^{\Lambda} f_1(\Lambda - U) f_2(U)\, dU. \qquad (5.1.27)$$

Here the limits mean that the integral is over all $U \in S_m$ with $0 < U < \Lambda$.

But it should be remembered that in general $\int_\Lambda^M + \int_M^N \neq \int_\Lambda^N$, if $m(m+1)/2 > 1$.

As an application of (5.1.27), we can define the beta integral as follows:

Definition 5.1.5 *The Generalized Beta Function*

$$B_m(\alpha, \beta) = \int_0^I (\det U)^{\alpha-(m+1)/2} \det(I - U)^{\beta-(m+1)/2}\, dU \qquad (5.1.28)$$

for $R(\alpha), R(\beta) > (m+1)/2 - 1$ where I is an identity matrix. By making a transformation $U \to \Lambda^{1/2} U \Lambda^{1/2}$, we get

$$B_m(\alpha, \beta)(\det \Lambda)^{\alpha+\beta-(m+1)/2} = \int_0^{\Lambda} (\det U)^{\alpha-(m+1)/2} \det(\Lambda - U)^{\beta-(m+1)/2}\, dU \qquad (5.1.29)$$

Now by taking the Laplace transform, we get

$$B_m(\alpha, \beta)\, \Gamma_m(\alpha + \beta) = \Gamma_m(\alpha)\, \Gamma_m(\beta) \qquad (5.1.30)$$

We shall obtain a few results when the matrices are rectangular. Let $S_{k,m}$ be the space of all $k \times m$ matrices. That is, $T \in S_{k,m}$ implies that $T = (t_{\alpha i})$, $\alpha = 1, \ldots, k$, $i = 1, \ldots, m$ (k rows and m columns) with the volume element $dT = \prod_{\alpha=1, i=1}^{k, m} dt_{\alpha i}$. If T is transformed to $T \to TZ$ with Z a real nonsingular $m \times m$ matrix then $dT \to |\det Z|^k\, dT$.

If $k \geqslant m$ then for $T \in S_{k,m}$ we have $T'T = R \geqslant 0$ and we may write T uniquely in the form $T = VR^{1/2}$ where R is a positive definite $m \times m$ symmetric matrix and V a $k \times m$ matrix such that $V'V = I$ where I is the identity matrix, that is, V is an element of the Stieffel manifold $V_{k,m}$, the collection of all m-tuples of orthonormal k-vectors. In this case the volume element is the following: For all $T \in S_{k,m}$ with $T = VR^{1/2}$, $R > 0$, $V \in V_{k,m}$ we have

$$dT = \frac{(\det R)^{(k-m-1)/2}}{2^m}\, dR\, dV \qquad (5.1.31)$$

It is also easy to show that

$$\int_{V_{k,m}} dV = \frac{2^m \pi^{(m-k)/2}}{\Gamma_m\left(\frac{k}{2}\right)} = C_{k,m} \tag{5.1.32}$$

This can be proved by the following technique. By direct integration one can show that

$$\int_{S_{k,m}} \text{etr}\,(-T'T)\,dT = \pi^{mk/2} \tag{5.1.33}$$

Now by writing $T = VR^{1/2}$, $R > 0$, $V \in V_{k,m}$ we have

$$\pi^{mk/2} = \int_{S_{k,m}} \text{etr}\,(-T'T)\,dT \tag{5.1.34}$$

$$= 2^{-m} \int_{R>0} \text{etr}\,(-R)\,(\det R)^{(k-m-1)/2}\,dR \int_{V_{k,m}} dV$$

$$= 2^{-m}\,\Gamma_m\left(\frac{k}{2}\right) \int_{V_{k,m}} dV$$

The following are two important Fourier transforms.

For $\qquad R(Z) > 0,\ S,\ T \in S_{k,m}$

$$\int_{S_{k,m}} \text{etr}\,(-2\pi i\,S'T)\ \text{etr}\,(-\pi\,TZT')\,dT \tag{5.1.35}$$

$$= \text{etr}\,\{-\pi\,SZ^{-1}S'\}\,(\det Z)^{-k/2}$$

For Z a complex $m \times m$ matrix with $R(Z) > 0$ and $P(T)$ a H-polynomial of degree ν we have

$$\int_{S_{k,m}} \text{etr}\,(-2\pi i\,S'T)\ \text{etr}\,(-\pi T'ZT)\,P(T)\,dT \tag{5.1.36}$$

$$= (\det Z)^{-k/2-\nu}\,\text{etr}\,(-\pi\,S'Z^{-1}S)\,P(-iS).$$

DEFINITION 5.1.6 *H-Polynomials of Degree* ν. Let T be a $k \times m$ matrix and $P(T)$ be a polynomial in the elements of T such that,

(1) $P(T)$ is harmonic, that is,

$$\sum_{\alpha=1,\,i=1}^{k,\,m} \frac{\partial^2 P}{\partial\,(t_{\alpha i})^2} = 0$$

and

(2) $P(TZ) = (\det Z)^\nu P(T)$ for all $m \times m$ matrices Z. Then $P(T)$ is called a H-polynomial of degree ν.

H-polynomials have the property that they are annihilated by each of the operators $\Delta_{ij} = \sum_{\alpha=1}^{k} \partial^2/\partial t_{\alpha i}\, \partial t_{\alpha j}$, individually. That is,

$$\Delta_{ij} f(T) = 0 \tag{5.1.37}$$

A fundamental solution of the differential equation (5.1.37) can be easily seen to be,

$$f(T) = \det (T'T)^{-(k-m-1)/2}. \tag{5.1.38}$$

5.2 SOME ELEMENTARY FUNCTIONS OF MATRIX ARGUMENT

In Section 5.1 we have defined the gamma and beta functions. In this section we will start with the definition of a generalized hypergeometric series of matrix argument and then list the various elementary functions which are its special cases. Integral representations are also given in the case of elementary functions.

DEFINITION 5.2.1 *Hypergeometric Functions of Matrix Argument.*

$${}_pF_q\,(a_1, \ldots, a_p; b_1, \ldots, b_q; Z) = \sum_{k=0}^{\infty} \sum_{K} \frac{(a_1)_K \ldots (a_p)_K\, C_K(Z)}{(b_1)_K \ldots (b_q)_K\, k!} \tag{5.2.1}$$

where

$$(a)_K = \prod_{i=1}^{m} \left(a - \frac{1}{2}(i-1)\right)_{k_i}, \quad K = (k_1, \ldots, k_m) \tag{5.2.2}$$

$$(a)_k = a(a+1) \ldots (a+k-1), \quad (a)_0 = 1.$$

If the a's and b's are such that the gamma functions are defined, then,

$$(a)_K = \frac{\Gamma_m(a, K)}{\Gamma_m(a)} \tag{5.2.3}$$

where

$$\Gamma_m(a, K) = \pi^{m(m-1)/4} \prod_{i=1}^{m} \Gamma\left(a + k_i - \frac{1}{2}(i-1)\right) \tag{5.2.4}$$

and

$$\Gamma_m(a) = \pi^{m(m-1)/4} \prod_{i=1}^{m} \Gamma\left(a - \frac{1}{2}(i-1)\right). \tag{5.2.5}$$

Z is a complex $m \times m$ symmetric matrix and $C_K(Z)$ are the zonal polynomials given in Definition 5.1.2. The series (5.2.1) is defined for $p \leqslant q+1$, and when $p = q+1$ the series converges for $\|Z\| < 1$ where

$\|Z\|$ denotes the largest of the absolute values of the characteristic roots of Z. The series converges for all Z when $p \leqslant q$. The parameters a_j's and b_j's are complex numbers none of which is an integer or a half integer $\leqslant (m-1)/2$.

Usually series representations have many disadvantages and hence we may define the hypergeometric functions by an inductive process, starting with the definition of a gamma function. By an inductive process we can give a definition in the following integral form.

$$\begin{aligned} &{}_{p+1}F_q(a_1,\ldots,a_p,\gamma;b_1,\ldots,b_q,-Z^{-1})(\det Z)^{-\gamma} \qquad (5.2.6)\\ &=\frac{1}{\Gamma_m(\gamma)}\int\limits_{\Lambda>0} \operatorname{etr}(-\Lambda Z)\,{}_pF_q(a_1,\ldots,a_p;b_1,\ldots,b_q;-\Lambda)\\ &\qquad\times(\det\Lambda)^{\gamma-(m+1)/2}\,d\Lambda \end{aligned}$$

for $R(Z)>0$, $R(\gamma)>(m+1)/2-1$. By a simple change of variables one can write, for $Z>0$,

$$\begin{aligned} &{}_{p+1}F_q(a_1,\ldots,a_p,\gamma;b_1,\ldots,b_q;Z) \qquad (5.2.7)\\ &=\frac{1}{\Gamma_m(\gamma)}\int\limits_{\Lambda>0} \operatorname{etr}(-\Lambda)\,{}_pF_q(a_1,\ldots,a_p;b_1,\ldots,b_q;\Lambda Z)\\ &\qquad\times(\det\Lambda)^{\gamma-(m+1)/2}\,d\Lambda. \end{aligned}$$

We can also define it in terms of an inverse-Laplace transform as follows:

$$\begin{aligned} &{}_pF_{q+1}(a_1,\ldots,a_p;b_1,\ldots,b_q,\gamma;-\Lambda)(\det\Lambda)^{\gamma-(m+1)/2} \qquad (5.2.8)\\ &=\frac{\Gamma_m(\gamma)}{(2\pi i)^{m(m+1)/2}}\int\limits_{R(Z)=X_0>0} \operatorname{etr}(\Lambda Z)\,{}_pF_q(a_1,\ldots,a_p;\\ &\qquad\qquad b_1,\ldots,b_q;-Z^{-1})(\det Z)^{-\gamma}\,dZ \end{aligned}$$

for $\Lambda>0$, $R(\gamma)>(m+1)/2-1$. We may also write it as,

$$\begin{aligned} &{}_pF_{q+1}(a_1,\ldots,a_p;b_1,\ldots,b_q,\gamma;\Lambda) \qquad (5.2.9)\\ &=\frac{\Gamma_m(\gamma)}{(2\pi i)^{m(m+1)/2}}\int\limits_{R(Z)=X_0>0} \operatorname{etr}(Z)\\ &\times{}_pF_q(a_1,\ldots,a_p;b_1,\ldots,b_q;\Lambda Z^{-1})(\det Z)^{-\gamma}\,dZ. \end{aligned}$$

Now we will list the special cases of (5.2.1)

DEFINITION 5.2.2 *Generalized Exponential Series.*

$$e^{\operatorname{tr} Z}=\operatorname{etr} Z={}_0F_0(Z)=\sum_{k=0}^{\infty}\sum_{K}\frac{C_K(Z)}{k!}. \qquad (5.2.10)$$

DEFINITION 5.2.3 *Generalized Binomial Series.*

$$ {}_1F_0(a; Z) = \det(I - Z)^{-a} = \sum_{k=0}^{\infty} \sum_{K} (a)_K \frac{C_K(Z)}{k!} \qquad (5.2.11)$$

$$= \frac{1}{\Gamma_m(a)} \int_{\Lambda>0} \operatorname{etr}(-\Lambda) \operatorname{etr}(\Lambda Z) (\det \Lambda)^{a-(m+1)/2} \, d\Lambda$$

for $R(Z) < I$ and $R(a) > (m+1)/2 - 1$. It is analytic for all a and $R(Z) < I$.

DEFINITION 5.2.4 *Bessel Functions of Matrix Argument.*

$$ {}_0F_1(\gamma; Z) = \frac{\Gamma_m(\gamma)}{(2\pi i)^{m(m+1)/2}} \int_{R(\Lambda)=X_0>0} \operatorname{etr}(\Lambda) \operatorname{etr}(-\Lambda^{-1} Z) (\det \Lambda)^{-\gamma} \, d\Lambda \quad (5.2.12)$$

$$= \sum_{k=0}^{\infty} \sum_{K} \frac{C_K(Z)}{(\gamma)_K \, k!}$$

for $$R(\gamma) > \frac{m+1}{2} - 1.$$

A few interesting properties of this function are worth recording. Let

$$A_\delta(Z) = \frac{1}{\Gamma_m\left(\delta + \frac{m+1}{2}\right)} \, {}_0F_1\left(\delta + \frac{m+1}{2}; -Z\right) \qquad (5.2.13)$$

then

$$A_\delta(M) = \frac{1}{(2\pi i)^{m(m+1)/2}} \int_{R(Z)=X_0>0} \operatorname{etr}(Z) \operatorname{etr}(-MZ^{-1}) \times (\det Z)^{-\delta-(m+1)/2} \, dZ \qquad (5.2.14)$$

for $$R(\delta) > \frac{m+1}{2} - 1.$$

$$D\{(\det \Lambda)^{\delta} A_\delta(\Lambda)\} = (\det \Lambda)^{\delta-1} A_{\delta-1}(\Lambda) \qquad (5.2.15)$$

and

$$D\{A_\delta(M)\} = (-1)^{\delta} A_{\delta+1}(M) \qquad (5.2.16)$$

where D is the differential operator defined in (5.1.5).

DEFINITION 5.2.5 *Confluent Hypergeometric Functions of Matrix Argument.*

$$ {}_1F_1(\alpha; \beta; Z) = \frac{\Gamma_m(\beta)}{(2\pi i)^{m(m+1)/2}} \int_{R(\Lambda)=X_0>0} \operatorname{etr}(\Lambda) \times \det(I - Z\Lambda^{-1})^{-\alpha} (\det \Lambda)^{-\beta} \, d\Lambda \qquad (5.2.17)$$

for

$$X_0 > R(Z) \text{ and } R(\beta) > m.$$

$$= \frac{\Gamma_m(\beta)}{\Gamma_m(\alpha)\,\Gamma_m(\beta-\alpha)} \int_0^I \operatorname{etr}(\Lambda Z)\,(\det \Lambda)^{\alpha-(m+1)/2} \times \det(I-\Lambda)^{\beta-\alpha-(m+1)/2}\, d\Lambda \qquad (5.2.18)$$

for

$$R(\alpha) > \frac{m+1}{2} - 1,\ R(\beta) > \frac{m+1}{2} - 1$$

and

$$R(\beta-\alpha) > \frac{m+1}{2} - 1$$

$$= \sum_{k=0}^{\infty} \sum_{K} \frac{(\alpha)_K}{(\beta)_K} \frac{C_K(Z)}{k!} \qquad (5.2.19)$$

By the uniqueness of Laplace transform we get Kummer's formula

$$\operatorname{etr}(-Z)\,{}_1F_1(\alpha;\beta;Z) = {}_1F_1(\beta-\alpha;\beta;-Z). \qquad (5.2.20)$$

By a change of variable in (5.2.18) we have

$${}_1F_1(\alpha;\beta;-\Lambda) = \frac{\Gamma_m(\beta)\,(\det \Lambda)^{(m+1)/2-\beta}}{\Gamma_m(\alpha)\,\Gamma_m(\beta-\alpha)}$$

$$\times \int_0^{\Lambda} \operatorname{etr}(-M)\,(\det M)^{\alpha-(m+1)/2}\,(\det(\Lambda-M))^{\beta-\alpha-(m+1)/2}\, dM \qquad (5.2.21)$$

for $\Lambda > 0$.

DEFINITION 5.2.6 *Laguerre Functions of Matrix Argument.*

$$L_\nu^{(\gamma)}(Z) = \frac{\Pi_m(\gamma+\nu)}{\Pi_m(\gamma)}\,{}_1F_1\left(-\nu, \gamma + \frac{m+1}{2}; Z\right) \qquad (5.2.22)$$

where for example

$$\Pi_m(a) = \Gamma_m\left(a + \frac{m+1}{2}\right),$$

for

$R(\gamma) > -1,\ R(\gamma+\nu) > -1.$ Also for $R(\gamma) > (m+1)/2,\ Z > 0$

we have

$$L_\nu^{(\gamma)}(Z) = \frac{\operatorname{etr}(Z)\,(\det Z)^{-\gamma}}{\Gamma_m(-\nu)} \int_0^Z \operatorname{etr}(-\Lambda)\,(\det \Lambda)^{\gamma+\nu} \times (\det(Z-\Lambda))^{-\nu-(m+1)/2}\, d\Lambda. \qquad (5.2.23)$$

When $R(\gamma+\nu) > -1$ and $R(Z) > 0$ we may write

$$L_\nu^{(\gamma)}(Z) = \operatorname{etr}(Z)(\det Z)^{-\gamma} D^\nu \{\operatorname{etr}(-Z)(\det Z)^{\gamma+\nu}\} \tag{5.2.24}$$

and when ν is a nonnegative integer this generalizes Rodrigues' formula for Laguerre polynomials.

DEFINITION 5.2.7 *Gaussian Hypergeometric Functions of Matrix Argument.*

$$ {}_2F_1(\alpha, \beta; \gamma; Z) \tag{5.2.25}$$

$$= \frac{\Gamma_m(\gamma)}{\Gamma_m(\alpha)\Gamma_m(\gamma-\alpha)} \int_0^I \det(I-\Lambda Z)^{-\beta} (\det \Lambda)^{\alpha-(m+1)/2} \times \det(I-\Lambda)^{\gamma-\alpha-(m+1)/2}\, d\Lambda$$

for $R(\alpha) > \frac{m+1}{2} - 1$, $R(\gamma-\alpha) > \frac{m+1}{2} - 1$, all β and $R(\Lambda) < I$

$$= \sum_{k=0}^{\infty} \sum_{K} \frac{(\alpha)_K (\beta)_K}{(\gamma)_K} \frac{C_K(Z)}{k!} \tag{5.2.26}$$

A representation through inverse-Laplace transform is available from (5.2.8) and (5.2.9). Also Kummer's relations follow without much difficulty. That is,

$$\lim_{\beta\to\infty} {}_2F_1\left(\alpha, \beta; \gamma; \frac{Z}{\beta}\right) = {}_1F_1(\alpha; \gamma; -Z) \tag{5.2.27}$$

$$\lim_{\alpha\to\infty} {}_1F_1\left(\alpha; \beta; \frac{Z}{\alpha}\right) = {}_0F_1(\beta; -Z). \tag{5.2.28}$$

We can also derive Euler's formulae

$${}_2F_1(\alpha, \beta; \gamma; Z) = \det(I-Z)^{-\beta}\, {}_2F_1(\gamma-\alpha, \beta; \gamma; -Z(I-Z)^{-1}) \tag{5.2.29}$$

for $R(\gamma) > \frac{m+1}{2} - 1$, $R(\gamma-\alpha) > \frac{m+1}{2} - 1$, $R(Z) < I$ and

$${}_2F_1(\alpha, \beta; \gamma; Z) = \det(I-Z)^{\gamma-\alpha-\beta}\, {}_2F_1(\gamma-\alpha, \gamma-\beta; \gamma; Z) \tag{5.2.30}$$

For $R(\gamma+\nu) > -1$ and $0 < R(Z) < I$ we have

$${}_2F_1\left(\gamma+\nu+\frac{m+1}{2}, -\beta; \gamma+\frac{m+1}{2}; Z\right) = \frac{\Gamma_m\left(\gamma+\frac{m+1}{2}\right)}{\Gamma_m\left(\gamma+\nu+\frac{m+1}{2}\right)} \times (\det Z)^{-\gamma} D^\nu \{(\det Z)^{\gamma+\nu} \det(I-Z)^{\beta}\}. \tag{5.2.31}$$

DEFINITION 5.2.8 *Jacobi Polynomials of Matrix Argument.*

$$P_\nu^{(\gamma,\delta)}(\Lambda) = \frac{\Gamma_m\left(\gamma+\nu+\frac{m+1}{2}\right)}{\Gamma_m\left(\gamma+\frac{m+1}{2}\right)} \times {}_2F_1\left(-\nu, \gamma+\delta+\nu+\frac{m+1}{2}; \gamma+\frac{m+1}{2}; \Lambda\right) \qquad (5.2.32)$$

for $0 < R(\Lambda) < I$ and $R(\gamma) > -1$. By using (5.2.30), we have

$$P_\nu^{(\gamma,\delta)}(\Lambda) = \frac{\Gamma_m\left(\gamma+\nu+\frac{m+1}{2}\right)}{\Gamma_m\left(\gamma+\frac{m+1}{2}\right)} \det(I-\Lambda)^{-\delta} \times {}_2F_1\left(\gamma+\nu+\frac{m+1}{2}, -\delta-\nu; \gamma+\frac{m+1}{2}; \Lambda\right) \qquad (5.2.33)$$

Now by using (5.2.31) we get

$$P_\nu^{(\gamma,\delta)}(\Lambda) = (\det\Lambda)^{-\gamma}\det(I-\Lambda)^{-\delta} D^\nu\{(\det\Lambda)^{\gamma+\nu}\det(I-\Lambda)^{\delta+\nu}\} \qquad (5.2.34)$$

For ν a nonnegative integer, $P_\nu^{(\gamma,\delta)}(\Lambda)$ is a symmetric polynomial of degree $m\nu$. For the weight function $(\det\Lambda)^\gamma \det(I-\Lambda)^\delta$ on the interval $0 < \Lambda < I$, $P_\nu^{(\gamma,\delta)}(\Lambda)$ is orthogonal to every polynomial of degree less than $m\nu$. When $m = 1$ these polynomials are the classical Jacobi polynomials transferred to the interval [0, 1].

DEFINITION 5.2.9 *Gegenbauer Polynomials of Matrix Argument.* Let T be a $m \times m$ nonsingular symmetric matrix. For $R(\delta) > m/2 - 1$ the Gegenbauer polynomials may be defined by the formulae,

$$C_{2\nu}^{(\delta)}(T) = (-1)^{m\nu}\frac{\Gamma_m(\delta+\frac{1}{2})}{\Gamma_m(\delta+\nu+\frac{1}{2})} P_\nu^{(-\frac{1}{2},\delta-m/2)}(T'T) \qquad (5.2.35)$$

and

$$C_{2\nu+1}^{(\delta)}(T) = (-1)^{m\nu}\frac{\Gamma_m(\delta+\frac{1}{2})}{\Gamma_m(\delta+\nu+\frac{1}{2})} P_\nu^{(\frac{1}{2},\delta-m/2)}(T'T)(\det T) \qquad (5.2.36)$$

These polynomials $C_\nu^{(\delta)}$ are orthogonal on $T'T < I$ with the weight function $\det(I - T'T)^{\delta-m/2}$. When $\delta = m/2$, we have the Legendre case.

DEFINITION 5.2.10 *Generalized Sine and Cosine.* These are defined for an $m \times m$ matrix Z as the average of complex exponentials over the full orthogonal group as follows:

$$\cos_m Z = \frac{1}{C_{m,m}} \int_{0(m)} \operatorname{etr}(i0Z)\, d0, \tag{5.2.37}$$

and

$$\sin_m Z = \frac{1}{i^m} \frac{1}{C_{m,m}} \int_{0(m)} \operatorname{etr}(i0S)\,(\det 0)\, d0$$

where $C_{k,m}$ is defined in (5.1.32). It can be shown that

$$\cos_m Z = \Gamma_m\left(-\frac{1}{2} + \frac{m+1}{2}\right) A^{(m)}_{-1/2}\left(\frac{Z'Z}{4}\right) \tag{5.2.38}$$

and

$$\sin_m Z = \frac{\Gamma_m\left(-\frac{1}{2} + \frac{m+1}{2}\right)(\det Z)}{2^m} A_{1/2}\left(\frac{Z'Z}{4}\right)$$

where

$$A^{(m)}_{\gamma}(Z)(\det Z)^{\nu/2} = \frac{1}{\Gamma_m\left(\frac{K}{2}\right) i^{m\nu}} \frac{1}{C_{k,m}} \int_{V_{k,m}} \operatorname{etr}(2iVZ^{1/2})\, P(V)\, dV \tag{5.2.39}$$

with $\gamma = \nu + (k - m - 1)/2$ and $P(V)$ is a H-polynomial of degree ν such that $P(E_{k,m}) = 1$ where $E_{k,m}$ denotes the $k \times m$ matrix with 1's on the leading diagonal and zeros elsewhere. The space $V_{k,m}$ is the collection of all m-tuples of orthonormal k-vectors.

DEFINITION 5.2.11 *Meijer's G-functions of Matrix Argument.* Meijer's G-function is the generalization of the generalized hypergeometric function. A detailed discussion of this function, when the argument is scalar, is available in Mathai and Saxena (1973a). Here we will extend the definition to the case of matrix argument. Mathai and Saxena (1971a) used a series representation to define G-functions of matrix argument. Since this definition is restricted to the conditions of expansibility we will define the G-function as an inverse matrix transform following Mathai (1976). Consider the function

$$\int_{Z>0} (\det Z)^{\rho-(m+1)/2}\, G^{p,q}_{r,s}\left[Z \,\middle|\, \begin{matrix} a_1, \ldots, a_r \\ b_1, \ldots, b_s \end{matrix} \right] dZ \tag{5.2.40}$$

$$= \frac{\prod_{j=1}^{p} \Gamma_m(b_j + \rho) \prod_{j=1}^{q} \Gamma_m\left(\frac{m+1}{2} - a_j - \rho\right)}{\prod_{j=p+1}^{s} \Gamma_m\left(\frac{m+1}{2} - b_j - \rho\right) \prod_{j=q+1}^{r} \Gamma_m(a_j + \rho)}$$

where Z is a $m \times m$ matrix, $\Gamma_m(\ )$ is defined in (5.1.2) and the integral is taken over all positive definite matrices Z. $R(b_j+\rho)>(m-1)/2$; $(j=1,\ldots,m)$ and $R(a_j+\rho)<(m+1)/2$ $(j=1,\ldots,n)$. The gamma products are such that the poles of $\prod_{j=1}^{p}\Gamma_m(b_j+\rho)$ and those of $\prod_{j=1}^{q}\Gamma_m[(m+1)/2-a_j-\rho]$ are separated, $q \geqslant 1$ and $p<q$ or $p=q$ and $0<Z<I$. That is, the G-function of the matrix argument is that function for which the matrix transform is the gamma product given on the R.H.S. of (5.2.40). The matrix transform of a function is defined as follows.

DEFINITION 5.2.12 *The Matrix Transform of $f(Z)$.* Let

$$\int_{Z>0} (\det Z)^{\rho-(m+1)/2} f(Z)\, dZ = g(\rho) \tag{5.2.41}$$

for $R(\rho)>(m+1)/2-1$. Then, whenever $g(\rho)$ exists, it will be called the matrix transform of $f(Z)$ where Z is a $m \times m$ matrix variable. By using the definition in (5.2.41) and by using the property that if $Z=U^{1/2}VU^{1/2}$ then $dZ=(\det U)^{(m+1)/2}\,dV$ we can obtain the multiplication formula for the matrix transforms.

Consider the class $\{f(Z)\}$ of symmetric functions of the symmetric positive definite matrix Z. Let $f_1(Z)$ and $f_2(Z)$ be two members of this class with matrix transforms $g_1(\rho)$ and $g_2(\rho)$ respectively. Then from Mathai (1976) it can be seen that the matrix transform of $f_3(Z)$ is $g_1(\rho)\, g_2[(m+1)/2+\beta-\rho]$ where,

$$f_3(Z)=\int_{V>0} (\det V)^{\beta} f_1(ZV)\, f_2(V)\, dV. \tag{5.2.42}$$

Now by using the definition in (5.2.41) and identifying the resulting gamma products we can write down some special cases of the G-function of matrix argument. Such results will be discussed in Section 5.3. For the sake of illustration we will list one result here. From the definition of the gamma function given in (5.1.20) it follows that

$$\int_{Z>0} (\det Z)^{a+\rho-(m+1)/2}\, e^{-\operatorname{tr} Z}\, dZ = \Gamma_m(a+\rho) \tag{5.2.43}$$

Hence by comparing (5.2.43) and (5.2.40), we get

$$G_{0,1}^{1,0}(Z \mid a) = (\det Z)^{a}\, e^{-\operatorname{tr} Z} \tag{5.2.44}$$

DEFINITION 5.2.13 *H-functions of Matrix Argument.* As an extension of (5.2.40) we may define a H-function of matrix argument by the following relation:

$$\int_{Z>0} (\det Z)^{\rho-(m+1)/2}\, H_{r,s}^{p,q}\left(Z \,\middle|\, \begin{matrix} (a_1,\alpha_1),\ldots,(a_r,\alpha_r) \\ (b_1,\beta_1),\ldots,(b_s,\beta_s) \end{matrix} \right) dZ \tag{5.2.45}$$

$$= \frac{\prod_{j=1}^{p} \Gamma_m(b_j+\beta_j\rho) \prod_{j=1}^{q} \Gamma_m\left(\frac{m+1}{2}-a_j-\alpha_j\rho\right)}{\prod_{j=p+1}^{s} \Gamma_m\left(\frac{m+1}{2}-b_j-\beta_j\rho\right) \prod_{j=q+1}^{r} \Gamma_m(a_j+\alpha_j\rho)}$$

where the integral is taken over all positive definite $m \times m$ matrices Z, $R(b_j+\rho\beta_j) > \left(\frac{m-1}{2}\right)\beta_j\ (j=1,\ldots,m)$ and $R\left(\frac{m+1}{2}-a_j-\rho\alpha_j\right) > \left(\frac{1-m}{2}\right)\alpha_j\ (j=1,\ldots,n)$; a_j and b_j are complex numbers and α_j and β_j are positive numbers such that the poles of the gammas in $\prod_{j=1}^{p} \Gamma_m(b_j+\beta_j\rho)$ and those of $\prod_{j=1}^{q} \Gamma_m\left(\frac{m+1}{2}-a_j-\alpha_j\rho\right)$ are separated. The parameters are such that the gamma products are defined.

Also from (5.2.40) and (5.2.41) we may define the hypergeometric function ${}_pF_q(a_1,\ldots,a_p;b_1,\ldots,b_q;Z)$ in terms of a matrix transform as follows. It is that function, whenever it exists, for which

$$\int_{Z>0} (\det Z)^{\rho-(m+1)/2}\, {}_pF_q(a_1\ldots,a_p;b_1,\ldots,b_q;Z)\, dZ \tag{5.2.46}$$

$$= \frac{\prod_{j=1}^{p} \Gamma_m(a_j-\rho)\,\Gamma_m(\rho)}{\prod_{j=1}^{q} \Gamma_m(b_j-\rho)} \frac{\prod_{j=1}^{q} \Gamma_m(b_j)}{\prod_{j=1}^{p} \Gamma_m(a_j)}$$

for $R(\rho) > (m+1)/2-1$, $R(a_j-\rho) > (1-m)/2\ (j=1,\ldots,p)$ and the parameters are such that the gamma products exist.

REMARK. It should be remarked that $f(Z)$ in (5.2.41) need not be unique. Since ρ is scalar we can not invert $g(\rho)$ to get $f(Z)$. Hence (5.2.41) is to be treated as a functional equation in $f(\cdot)$.

5.3 SOME INTEGRALS INVOLVING FUNCTIONS OF MATRIX ARGUMENT

In this section we will give some typical formulas involving special functions of matrix argument. Further results are available in the exercises at the end of this chapter.

Two of the basic integrals are available in the definitions of the gamma and beta functions. That is,

$$\int_{Z>0} (\det Z)^{\alpha-(m+1)/2}\, e^{-\operatorname{tr} Z}\, dZ = \Gamma_m(\alpha),\ R(\alpha) > \frac{m+1}{2} - 1 \tag{5.3.1}$$

$$\int_0^I (\det Z)^{\alpha-(m+1)/2}\, (\det(I-Z))^{\beta-(m+1)/2} = \frac{\Gamma_m(\alpha)\,\Gamma_m(\beta)}{\Gamma_m(\alpha+\beta)} \tag{5.3.2}$$

for $\qquad R(\alpha),\ R(\beta) > \dfrac{m+1}{2} - 1.$

THEOREM 5.3.1 *For a symmetric $m \times m$ positive definite matrix A and for*

$$R(\nu) > R(s) > \frac{m+1}{2} - 1,$$

$$\int_{Z>0} (\det Z)^{s-(m+1)/2}\, (\det(I+AZ))^{-\nu}\, dZ = (\det A)^{-s}\, \frac{\Gamma_m(s)\,\Gamma_m(\nu-s)}{\Gamma_m(\nu)} \tag{5.3.3}$$

where the integral is over all $m\times m$ symmetric positive definite matrices.

Proof. Since A is positive definite and $\det(I+AZ) = \det(I+ZA) = \det(I+A^{1/2}ZA^{1/2})$, where $A^{1/2}$ is uniquely defined, we may make the transformation $U = A^{1/2}ZA^{1/2}$, that is, $dU = (\det A)^{(m+1)/2}\, dZ$, to obtain,

$$\int_{Z>0} (\det Z)^{s-(m+1)/2} \det(I+AZ)^{-\nu}\, dZ \tag{5.3.4}$$

$$= (\det A)^{-s} \int_{U>0} (\det U)^{s-(m+1)/2}\, (\det(I+U))^{-\nu}\, dU$$

Let $V = (I+U)^{-1}$ then from (5.1.10) we have $dU = (\det V)^{-(m+1)}\, dV$. Also $U > 0$ implies that $I + U > I > 0$ or $I > (I+U)^{-1} > 0$. Hence we have,

$$\int_{U>0} (\det U)^{s-(m+1)/2} (\det(I-U))^{-\nu}\, dU \tag{5.3.5}$$

$$= \int_0^I (\det(V^{-1}-I))^{s-(m+1)/2}\, (\det V^{-1})^{-\nu}\, (\det V)^{-(m+1)}\, dV$$

$$= \int_0^I (\det V)^{\nu-s-(m+1)/2}\, (\det(I-V))^{s-(m+1)/2}\, dV.$$

Now by using (5.3.2) the result is established.

By using the results (5.2.40), (5.2.41) and (5.2.42) we can establish the following formulae for the G-functions of matrix argument.

$$\int_0^I (\det Z)^{\alpha-(m+1)/2} (\det (I-Z))^{\beta-(m+1)/2} G^{p,q}_{r,s}\left[RZ \,\middle|\, \begin{matrix} a_1,\ldots, a_r \\ b_1,\ldots, b_s \end{matrix}\right] dZ \tag{5.3.6}$$

$$= \Gamma_m(\beta)\, G^{p,q+1}_{r+1,s+1}\left[R \,\middle|\, \begin{matrix} \frac{m+1}{2}-\alpha, a_1,\ldots, a_p \\ b_1,\ldots, b_q, \frac{m+1}{2}-\alpha-\beta \end{matrix}\right]$$

for $R(\alpha + \min b_j)$ $(j = 1,\ldots, p)$, $R(s) > (m+1)/2 - 1$ and the existence conditions for the G-functions.

$$\int_{Z>0} (\det Z)^{-\alpha-(m+1)/2} e^{-\operatorname{tr} AZ} G^{p,q}_{r,s}\left[z \,\middle|\, \begin{matrix} a_1,\ldots, a_r \\ b_1,\ldots, b_s \end{matrix}\right] dZ \tag{5.3.7}$$

$$= (\det A)^{\alpha} G^{p,q+1}_{r+1,s}\left[A^{-1} \,\middle|\, \begin{matrix} \frac{m+1}{2}+\alpha, a_1,\ldots, a_r \\ b_1,\ldots, b_s \end{matrix}\right]$$

for $R(-\alpha + \min b_j) > (m+1)/2 - 1$ $(j = 1,\ldots, p)$ and A a positive definite symmetric matrix.

By specializing the parameters and noting from (5.2.40) and (5.2.46) that

$$ {}_pF_q (a_1, \ldots, a_p; b_1, \ldots, b_q: -Z) \tag{5.3.8}$$

$$= \frac{\prod_{j=1}^{q} \Gamma_m (b_j)}{\prod_{j=1}^{p} \Gamma_m(a_j)} G^{1,p}_{p,q+1}\left[Z \,\middle|\, \begin{matrix} \frac{m+1}{2} - a_1,\ldots, \frac{m+1}{2} - a_p \\ 0, \frac{m+1}{2} - b_1,\ldots, \frac{m+1}{2} - b_q \end{matrix}\right]$$

we obtain the following formulae from (5.3.6) and (5.3.7):

$$\int_0^I (\det Z)^{\alpha-(m+1)/2} (\det (I-Z))^{\beta-(m+1)/2} \times {}_rF_s (a_1,\ldots, a_r; b_1,\ldots, b_s; -AZ)\, dZ \tag{5.3.9}$$

$$= \frac{\Gamma_m (\alpha)\, \Gamma_m (\beta)}{\Gamma_m (\alpha+\beta)} {}_{r+1}F_{s+1} (\alpha, a_1,\ldots, a_r; \alpha+\beta, b_1,\ldots, b_s; A)$$

for $R(\alpha)$, $R(\beta) > (m+1)/2 - 1$, $r < s+1$, or $r = s+1$ and $0 < R < I$, A positive definite.

$$\int_{Z>0} (\det Z)^{\alpha-(m+1)/2} e^{-\operatorname{tr} ZA} {}_pF_q(a_1,\ldots,a_p;\ b_1,\ldots,b_q;\ -Z)\, dZ \tag{5.3.10}$$

$$= \Gamma_m(\alpha)\,(\det A)^{-\alpha}\ {}_{p+1}F_q\,(a_1,\ldots,a_p,\alpha;\ b_1,\ldots,b_q;\ -A^{-1})$$

for $R(A) > 0$, $R(\alpha) > (m+1)/2 - 1$.

Several interesting particular cases of the integrals (5.3.9) and (5.3.10) are given in the exercises at the end of this chapter. Some applications of these particular cases are mentioned in Section 5.4. Some of these particular cases will be used in establishing the following results. First, we will consider a result analogous to the incomplete gamma function in the univariate case. When the argument is a matrix we will define the incomplete gamma function as follows:

DEFINITION 5.3.1 *Incomplete Gamma Function.*

$$\gamma(\rho, A) = \int_0^A (\det Z)^{\rho-(m+1)/2} e^{-\operatorname{tr} Z}\, dZ \tag{5.3.11}$$

for $R(\rho) > (m+1)/2 - 1$.

THEOREM 5.3.2 *For* $R(\rho) > (m+1)/2 - 1$,

$$\gamma(\rho, A) = (\det A)^{\rho} \frac{\Gamma_m(\rho)\,\Gamma_m\left(\dfrac{m+1}{2}\right)}{\Gamma_m\left(\rho+\dfrac{m+1}{2}\right)}\ {}_1F_1\left(\rho;\ \rho+\frac{m+1}{2};\ -A\right) \tag{5.3.12}$$

Proof. By making the transformation $U = A^{-1/2} Z A^{-1/2}$, we get

$$\int_0^A (\det Z)^{\rho-(m+1)/2} e^{-\operatorname{tr} Z}\, dZ = (\det A)^{\rho} \int_0^I (\det U)^{\rho-(m+1)/2} e^{-\operatorname{tr} AU}\, dU \tag{5.3.13}$$

Now the result follows from (5.3.9) observing that ${}_0F_0\,(\cdot)$ is an exponential function.

In (5.3.13) the integral is over the region $0 < Z < A$ or $A - Z$ positive definite. It should be remembered that $Z > A$ gives the region where $A - Z$ is negative definite. But these two regions $Z \geqslant A$ and $Z < A$ do not complement each other because a matrix could be none of the following types: positive definite, negative definite and semi-definite. Thus when we are dealing with functions of matrix argument,

$$\int_A^B + \int_B^C \neq \int_A^C \tag{5.3.14}$$

Henee,

$$\int_{Z>A} (\det Z)^{\rho-(m+1)/2}\, e^{-\operatorname{tr} Z}\, dZ \neq \int_{Z>0} (\det Z)^{\rho-(m+1)/2}\, e^{-\operatorname{tr} Z}$$

$$-\int_0^A (\det Z)^{\rho-(m+1)/2}\, e^{-\operatorname{tr} Z}\, dZ \qquad (5.3.15)$$

That is,

$$\int_{Z>A} (\det Z)^{\rho-(m+1)/2}\, e^{-\operatorname{tr} Z}\, dZ \neq \Gamma_m(\rho) - \gamma(\rho, A). \qquad (5.3.16)$$

THEOREM 5.3.3 *For B $m \times m$ symmetric positive definite,*

$$R(\gamma - \rho) > \frac{m+1}{2} - 1,\ R(A^{-1}B^{-1}) < I,$$

$$\int_{Z>B} (\det Z)^{\rho-(k+1)/2} (\det(I + AZ))^{-\nu}\, dZ$$

$$= \frac{\Gamma_m(\nu-\rho)\,\Gamma_m\left(\frac{m+1}{2}\right)}{\Gamma_m\left(\nu - s + \frac{m+1}{2}\right)} (\det A)^{-\nu} (\det B)^{\rho-\nu} \qquad (5.3.17)$$

$$\times\, {}_2F_1\left(\nu-\rho,\ \nu;\ \nu-\rho+\frac{m+1}{2};\ -A^{-1}B^{-1}\right),$$

The proof follows from the following steps. Put $U = Z^{-1} \Rightarrow dZ = (\det U)^{-(m+1)}\, dU$, then put $BU = V \Rightarrow dV = (\det B)^{(m+1)/2}\, dU$. The respective regions are $0 < B < Z$, $0 < U < B^{-1}$, $0 < V < I$. Now with the help of (5.2.25) the result is established.

With the help of (5.2.25) we can also establish a representation for the incomplete beta function in terms of a ${}_2F_1(\cdot)$.

DEFINITION 5.3.2 *Incomplete Beta Function.* It is defined as the integral

$$\int_0^A (\det Z)^{\alpha-(m+1)/2} (\det(I - Z))^{\beta-(m+1)/2}\, dZ$$

which can be shown to be the following:

$$\int_0^A (\det Z)^{\alpha-(m+1)/2} (\det(I - Z))^{\beta-(m+1)/2}\, dZ \qquad (5.3.18)$$

$$= \frac{\Gamma_m(\alpha)\,\Gamma_m\left(\frac{m+1}{2}\right)}{\Gamma_m\left(\alpha + \frac{m+1}{2}\right)} (\det Z)^{\alpha}\, {}_2F_1\left(\alpha,\ -\beta + \frac{m+1}{2};\ \alpha + \frac{m+1}{2};\ A\right)$$

for $$R(\alpha) > (m+1)/2 - 1, \ 0 < A < I.$$

Before concluding this section we will discuss a few integrals involving zonal polynomials. These integrals will be useful when the zonal polynomials are applied to practical problems. Only the outlines of the proofs are given here. The details may be found in Constantine (1963). Zonal polynomials $C_K(Z)$ were defined in Section 5.1.

THEOREM 5.3.4 *Let Z be a complex symmetric matrix whose real part is positive definite and let T be an arbitrary complex symmetric matrix. Then*

$$\int_{S>0} e^{-\operatorname{tr} ZS} (\det S)^{\rho-(m+1)/2} C_K(ST)\, dS = (\det Z)^{-\rho} C_K(TZ^{-1}) \Gamma_m(\rho, K) \tag{5.3.19}$$

for $$R(\rho) > (m+1)/2 - 1$$

where

$$\Gamma_m(\rho, K) = \pi^{m(m-1)/4} \prod_{i=1}^{m} \Gamma_m\left(\rho + k_i - \frac{i-1}{2}\right) \tag{5.3.20}$$

and

$$K = (k_1, k_2, \ldots, k_m), \ k_1 \geqslant k_2 \geqslant \ldots \geqslant k_m \geqslant 0, \ k_1 + k_2 + \ldots + k_m = k.$$

Proof. When T is an identity matrix let

$$f(T) = \int_{S>0} e^{-\operatorname{tr} S} (\det S)^{\rho-(m+1)/2} C_K(ST)\, dS \tag{5.3.21}$$

$f(T)$ is a homogeneous symmetric polynomial. Making the transformation $T \to H'TH$ and integrating H over the orthogonal group $0(m)$ and using (5.1.17) we get

$$f(T) = \frac{f(I)\, C_K(T)}{C_K(I)} \text{ or } \Gamma_m(\rho, K) = \frac{f(I)}{C_K(I)}. \tag{5.3.22}$$

Assuming that T is diagonal and comparing coefficients of $t_1^{k_1} \ldots t_m^{k_m}$ on both sides of (5.3.21) we get

$$\Gamma_m(\rho, K) = \int_{S>0} e^{-\operatorname{tr} S} (\det S)^{\rho-(m+1)/2} (\det S_1)^{k_1-k_2} (\det S_2)^{k_2-k_3} \ldots (\det S_m)^{k_m}\, dS \tag{5.3.23}$$

Now put $S = U'U$ where U is an upper triangular matrix $U = (u_{ij})$, $u_{ij} = 0$ if $i > j$, $u_{ii} > 0$, $(\det S_k) = u_{11}^2 \ldots u_{kk}^2$ and the Jacobian of the transformation is available from (5.1.8) as $2^m \prod_{i=1}^{m} u_{ii}^{m-i+1}$. Hence

$$\Gamma_m(\rho, K) = \int \cdots \int e^{\left(\sum_{i<j} u_{ij}^2\right)} \prod_{i=1}^{m} (u_{ii}^2)^{\rho+k_i-(i+1)/2} \tag{5.3.24}$$

$$\times \prod_{i=1}^{m} du_{ii}^2 \prod_{i<j} du_{ij}$$

$$= \pi^{m(m-1)/4} \prod_{i=1}^{m} \Gamma\left(\rho + k_i - \frac{i-1}{2}\right)$$

This proves the result for $Z = I$. For the general case make the transformation $S \to Z^{1/2} S Z^{1/2}$.

THEOREM 5.3.5 *For Z a positive definite $m \times m$ matrix*

$$\int_0^I (\det S)^{\rho-(m+1)/2} (\det (I-S))^{\nu-(m+1)/2} C_K(ZS)\, dS \tag{5.3.25}$$

$$= \frac{\Gamma_m(\rho, K)\, \Gamma_m(\nu)}{\Gamma_m(\rho+\nu, K)} C_K(Z)$$

where $\Gamma_m(\nu) = \pi^{m(m-1)/4} \prod_{i=1}^{m} \Gamma\left(\nu - \frac{i-1}{2}\right)$.

The proof follows by making a transformation $S = Z^{-1/2}\, TZ^{-1/2}$ and then using the convolution theorem for the Laplace transform remembering that (5.3.19) is a Laplace transform.

Theorems 5.3.4 and 5.3.5 are two basic results which enable us to solve a large variety of problems involving zonal polynomials and hypergeometric functions. A detailed discussion of zonal polynomials is available from Subrahmaniam (1974).

5.4 APPLICATIONS TO STATISTICS

Special functions of matrix argument are mainly applied in the area of statistical distributions and especially in the problem of nonnull distributions of test statistics when the population is multivariate normal. The distribution of a test statistic, when the hypothesis is not assumed to be true, is called the nonnull distribution of that test statistic.

A $m \times 1$ real random vector X is said to have a multivariate normal distribution $N_m(\mu, \Sigma)$ if X has the density function

$$f(X) = \frac{e^{-1/2(X-\mu)'\Sigma^{-1}(X-\mu)}}{(\det \Sigma)^{1/2}(2\pi)^{m/2}} \tag{5.4.1}$$

where μ is a $m \times 1$ vector and Σ a $m \times m$ symmetric positive definite matrix of parameters, $(X-\mu)'$ denotes the transpose of $(X-\mu)$. Here

$-\infty < x_i < \infty$, $-\infty < \mu_i < \infty$, $i = 1, \ldots, m$, $\Sigma > 0$ where the x_i's and μ_i's are the elements of X and μ respectively.

In the multinormal case, that is, when the density of X is of the form (5.4.1), it can be easily shown that $E(X) = \mu$ and $\text{cov}(X) = E(X - EX)(X - EX)' = \Sigma$ where E denotes the operator "mathematical expectation".

DEFINITION 5.4.1 *Mathematical Expectation.* If x is a scalar random variable with density function $f(x)$ and if $\psi(x)$ is a function of x then

$$E(\psi(x)) = \int_{-\infty}^{\infty} \psi(x) f(x)\, dx. \qquad (5.4.2)$$

$E(\psi(x))$ may not exist depending upon $\psi(x)$ and $f(x)$. The definition can be modified to cover discrete cases as well as vector and matrix variable cases.

From (5.4.2) one may notice that if $\psi(x) = e^{-tx}$ then (5.4.2) is the Laplace transform of $f(x)$ if $f(x)$ is zero in the interval $(-\infty, 0)$ and t is an arbitrary real parameter. If $\psi(x) = e^{-itx}$, $i = (-1)^{1/2}$, then (5.4.2) defines the Fourier and double Fourier transforms of $f(x)$. If $\psi(x) = x^{s-1}$ then (5.4.2) defines the Mellin transform of $f(x)$. Thus one may define the inverse transforms by inverting (5.4.2).

In this section we are mainly concerned with vector and matrix random variables connected with the normal density (5.4.1).

DEFINITION 5.4.2 *A Simple Random Sample.* A set of random variables $X_1, \ldots, X_n$ (scalar, vector or matrix) which are independently (in probability sense) and identically distributed with common density function $f(x)$ is said to be a simple random sample of size n from $f(x)$.

DEFINITION 5.4.3 *A Statistic.* A function $T(X_1, \ldots, X_n)$ of $X_1, \ldots, X_n$ is said to be *a* statistic if T has its own density function.

Tests of statistical hypotheses such as the hypothesis that $\mu = \mu_0$ (given) in a $N(\mu, \Sigma)$ are carried out by constructing test statistics based on some criteria such as the likelihood ratio principle. According to the likelihood ratio principle we construct

$$\lambda = \frac{\max L(X_1, \ldots, X_n \mid H_0)}{\max L(X_1, \ldots, X_n)},\ 0 < \lambda \leqslant 1 \qquad (5.4.3)$$

where L is the likelihood function or the joint density function of the random sample $X_1, \ldots, X_n$ at the observed sample points and, for example, $\max L(X_1, \ldots, X_n \mid H_0)$ means the maximum of L under the hypothesis H_0. The hypothesis is rejected with probability α if there exists a λ_0 such that the probability that $0 < \lambda < \lambda_0$ is α for every preassigned value α. This procedure comes from the motivation that when the null hypothesis is true we have $\lambda = 1$ and hence the hypothesis should be rejected if the observed value of λ is closer to zero.

The distribution of test statistics such as λ, when the hypothesis is assumed to be true, is called the null distribution and otherwise it is called the nonnull distribution. One of the topics where functions of matrix argument are applied is the topic of nonnull distributions of the λ-criterion of (5.4.3) for testing various types of hypotheses on the parameters μ and Σ in a $N_m(\mu, \Sigma)$ or on the corresponding parameters when there are several such normal distributions involved. A large number of such test statistics in the multivariate normal case are functions of a matrix random variable known as a Wishart matrix. This Wishart matrix A has the structure $A = \sum_{\alpha=1}^{n} Z_\alpha Z'_\alpha$ where $Z_1, Z_2, \ldots, Z_n$ are $m \times 1$ vectors which are independently and identically distributed as $N_m(0, \Sigma)$ and Z'_α denotes the transpose of Z_α. This means that each Z_α has the expected value $E(Z_\alpha) = 0$ and the covariance matrix $\operatorname{cov}(Z_\alpha) = \Sigma$. The density function of A is known as the Wishart density which is given by

$$f_1(A) = C\,|A|^{(n-m-1)/2} \exp(-\tfrac{1}{2}\operatorname{tr} \Sigma^{-1} A),\ A > 0,\ \Sigma > 0 \qquad (5.4.4)$$

where C is the normalizing constant given by

$$C^{-1} = 2^{nm/2}\,\pi^{m(m-1)/4}\,|\Sigma|^{n/2} \prod_{i=1}^{m} \Gamma\left(\frac{n+1-i}{2}\right). \qquad (5.4.5)$$

The derivation of this density is available in any standard text book on multivariate statistical analysis. The Wishart density of A with parameters $m, n,$ and Σ is often denoted by $W_m(A, \Sigma, n)$.

In $A = \sum_{\alpha=1}^{n} Z_\alpha Z'_\alpha$ if we assume that $E(Z_\alpha) = \mu$ and $\operatorname{cov}(Z_\alpha) = \Sigma$ where μ is not necessarily a null vector then the distribution of A is known as the noncentral Wishart distribution, noncentral in the sense that $E(Z_\alpha) = \mu \neq 0$. This distribution is a fundamental distribution in the problems of nonnull distributions of test statistics. It is derived by many authors and we will give a derivation of it based on Constantine (1963). The necessary tools have already been discussed in Section 5.1. We will state this in the form of a theorem.

THEOREM 5.4.1 *If the $m \times n$ matrix X has the density function*

$$f_2(X) = (2\pi\Sigma)^{-n/2}\, e^{-(1/2)\operatorname{tr}\Sigma^{-1}(X-M)(X-M)'} \qquad (5.4.6)$$

then $S = XX'$ has the density function

$$f_3(S) = \frac{e^{-\text{tr}\,\Omega}\,(\det S)^{(n-m-1)/2}}{\Gamma_m\left(\frac{n}{2}\right)\,(\det 2\Sigma)^{n/2}}\; e^{-\text{tr}\,\frac{1}{2}\Sigma^{-1}S}\; {}_0F_1\,(n/2, \tfrac{1}{2}\Sigma^{-1}\Omega S) \tag{5.4.7}$$

where $\Omega = MM'\,\Sigma^{-1}/2$ is known as the noncentrality parameter, n is often called the degrees of freedom of S and ${}_0F_1(\cdot)$ is a hypergeometric function of matrix argument defined in Section 5.2. This ${}_0F_1(\cdot)$ has the expansion

$${}_0F_1\left(\frac{n}{2}; \frac{1}{2}\,\Sigma^{-1}\Omega\, S\right) = \sum_{k=0}^{\infty} \sum_{K} \frac{C_K\,(\frac{1}{2}\,\Sigma^{-1}\Omega\, S)}{\left(\frac{n}{2}\right)\quad k!} \tag{5.4.8}$$

where $C_K(\cdot)$ are the zonal polynomials discussed in Sections 5.1 and 5.3.

Proof. Let $f(S)$ denote the density function of S and $g(Z)$ its Laplace transform, that is,

$$g(Z) = E\,[e^{-\text{tr}\,ZXX'}] = \int_X e^{-\text{tr}\,ZXX'}\, f_2(X)\, dX \tag{5.4.9}$$

where E denotes mathematical expectation. The integration in (5.4.9) may be carried out by using (5.1.35) and (5.1.36) which imply that

$$\int e^{-tr\,RXX'}\, e^{\text{tr}\,SX'}\, dX = \pi^{mn/2}\, e^{\text{tr}\,(R^{-1}SS)/4}\,(\det R)^{-n/2} \tag{5.4.10}$$

Hence

$$g(Z) = (\det 2\Sigma)^{-n/2}\, e^{-\text{tr}\,\Sigma^{-1}MM'/2}\, e^{\text{tr}\,(\Sigma^{-1}MM'\Sigma^{-1})(Z+\frac{1}{2}\Sigma^{-1})^{-1}/4} \times \det(Z + \tfrac{1}{2}\,\Sigma^{-1})^{n/2}. \tag{5.4.11}$$

Putting $W = Z + \Sigma^{-1}/2$ in (5.4.11) and inverting we get

$$\begin{aligned} f(S) &= \frac{2^{m(m-1)/2}}{(2\pi i)^{m(m+1)/2}} \int_{R(Z)>0} e^{\text{tr}\,SZ}\; g(Z)\, dZ \qquad (5.4.12)\\ &= (\det 2\Sigma)^{-n/2}\, e^{-\text{tr}\,\Omega}\, e^{-\text{tr}\,\Sigma^{-1}S/2}\\ &\quad\times \frac{2^{m(m-1)/2}}{(2\pi i)^{m(m+1)/2}} \int_{R(W)=W_0>0} e^{\text{tr}\,WS}\, e^{\text{tr}\,\frac{1}{2}\Sigma^{-1}\Omega W^{-1}}\,(\det W)^{-n/2}\, dW \end{aligned}$$

where

$\Omega = \dfrac{1}{2}\Sigma^{-1}\, MM'$. Since $e^{\text{tr}\,\frac{1}{2}\Sigma^{-1}\Omega W^{-1}}\; {}_0F_0\,(\frac{1}{2}\Sigma^{-1}\,\Omega\, W^{-1})$

by using the definition of a ${}_pF_q(\cdot)$ in terms of a Laplace transform we get,

$$\frac{2^{m(m-1)/2}}{(2\pi i)^{m(m+1)/2}} \int_{R(W)=W_0} e^{\text{tr}\,WS}\; e^{\text{tr}\,\frac{1}{2}\Sigma^{-1}\Omega W^{-1}}\,(\det W)^{-n/2}\, dW \tag{5.4.13}$$

$$= \frac{1}{\Gamma_m\left(\frac{n}{2}\right)}\ {}_0F_1\left(\frac{n}{2};\frac{1}{2}\Sigma^{-1}\Omega S\right)(\det S)^{n/2-(m+1)/2}.$$

This establishes the result that $f(S)$ is $f_3(S)$ given in (5.4.7). Also from (5.4.7) we get the moments of (det S), that is,

$$E(\det S)^{s-1} = \frac{\Gamma_m\left(s-1+\frac{n}{2}\right)}{\Gamma_m\left(\frac{n}{2}\right)}(\det 2\Sigma)^{s-1}\ {}_1F_1\left(-s+1;\ \frac{n}{2};\ -\Omega\right) \quad (5.1.14)$$

$$= \frac{\Gamma_m\left(s-1+\frac{n}{2}\right)}{\Gamma_m\left(\frac{n}{2}\right)}(\det 2\Sigma)^{s-1}e^{-\operatorname{tr}\Omega}\ {}_1F_1\left(\frac{n}{2}+s-1;\frac{n}{2};\ \Omega\right).$$

The nonnull distributions of several test statistics can be derived with the help of the noncentral Wishart distribution (5.4.7). As an illustration we will derive a few such distributions. Consider the $(s-1)$th moment of (det S) discussed in (5.4.14). This (det S) is known as the sample generalized variance being the determinant of the sample covariance matrix. In (5.4.14) we have already derived its moments with the help of a noncentral Wishart distribution. From (5.4.14) we can obtain the distribution of (det S) by inverting (5.4.14) treating it as a Mellin transform of the density function of (det S). This is worked out and given in computable form by Mathai (1972b).

5.4.1 The Exact Noncentral Distribution of the Generalized Variance

From (5.4.14) the tth moment of (det S)/(det 2Σ) is given by,

$$E\left\{\left[\frac{\det S}{\det 2\Sigma}\right]^t\right\} = \frac{\Gamma_m\left(t+\frac{n}{2}\right)}{\Gamma_m\left(\frac{n}{2}\right)}\ e^{-\operatorname{tr}\Omega}\ {}_1F_1\left(t+\frac{n}{2};\frac{n}{2};\ \Omega\right) \quad (5.4.15)$$

$$= \frac{e^{-\operatorname{tr}\Omega}}{\Gamma_m\left(\frac{n}{2}\right)}\sum_{k=0}^{\infty}\sum_{K}\frac{\Gamma_m\left(t+\frac{n}{2}\right)}{\left(\frac{n}{2}\right)_K k!}C_K(\Omega)\left(t+\frac{n}{2}\right)_K$$

Let $f(x)$ denote the density function of (det S)/(det 2Σ), then from the inverse Mellin transform we have

$$f(x)=\frac{e^{-\operatorname{tr}\Omega}}{\Gamma_m\left(\frac{n}{2}\right)}\sum_{k=0}^{\infty}\sum_{K}\frac{C_K(\Omega)}{k!\left(\frac{n}{2}\right)_K}\frac{1}{2\pi i}\int_{c-i\infty}^{c+i\infty}\Gamma_m\left(t+\frac{n}{2}\right)\left(t+\frac{n}{2}\right)_K x^{-t-1}\,dt \tag{5.4.16}$$

where $i=\sqrt{-1}$ and c is chosen suitably. But,

$$\Gamma_m\left(t+\frac{n}{2}\right)=\pi^{m(m-1)/4}\,\Gamma\left(t+\frac{n}{2}\right)\Gamma\left(t+\frac{n}{2}-\frac{1}{2}\right)\cdots\Gamma\left(t+\frac{n}{2}-\frac{m-1}{2}\right) \tag{5.4.17}$$

and

$$\left(t+\frac{n}{2}\right)_K=\left(t+\frac{n}{2}\right)_{k_1}\left(t+\frac{n}{2}-\frac{1}{2}\right)_{k_2}\cdots\left(t+\frac{n}{2}-\frac{m-1}{2}\right)_{k_m},$$

$$k_1\geqslant k_2\geqslant\ldots\geqslant k_m\geqslant 0,\ k_1+k_2+\ldots+k_m=k.$$

Hence by evaluating (5.4.16) with the help of a G-function, we get,

$$f(x)=\frac{\pi^{m(m-1)/4}\,e^{-\operatorname{tr}\Omega}}{\Gamma_m\left(\frac{n}{2}\right)}\sum_{k=0}^{\infty}\sum_{K}\frac{C_K(\Omega)}{k!\left(\frac{n}{2}\right)_K}\,x^{-1}$$

$$\times\, G^{m,0}_{0,m}\left[x\,\middle|\,\frac{n}{2}+k_1,\ \frac{n}{2}-\frac{1}{2}+k_2,\ \ldots,\ \frac{n}{2}-\frac{m-1}{2}+k_m\right],\ 0<x<\infty \tag{5.4.18}$$

The G-function appearing in (5.4.18) is the logarithmic case which can be evaluated by using the techniques discussed in Mathai and Saxena (1973a). Since the expression for the G-function in (5.4.18) is lengthy it is not given here. It is available from Mathai (1972b) in terms of a computable series involving powers of x and $\log x$.

REMARK 1. Computations of percentage points or other numerical values are possible from the expressions in Mathai (1972b) provided one has the various zonal polynomials. Explicit expressions for the zonal polynomials are available only for some special cases. This is a drawback of the method through zonal polynomials.

REMARK 2. The topic of nonnull distributions of test statistics is still a growing area with lots of unsolved problems. Some problems are given in the exercises at the end of this chapter and some references to the literature in this direction are available from the references cited at the end of this book.

5.4.2 The Exact Nonnull Distributions of a Collection of Test Statistics

Consider the independent matrix variates $X(m \times n_1)$ and $Y(m \times n_2)$, $m \leqslant n_i, i = 1, 2$ with the columns of X and Y distributed as $N_m(0, \Sigma_1)$ and $N_m(0, \Sigma_2)$ respectively. Then $S_1 = XX'$ and $S_2 = YY'$ are independent and have the Wishart distributions $W_m(n_i, \Sigma_i)$, $i = 1, 2$. Let $0 < f_1 < \ldots < f_m < \infty$, be the eigenvalues of $\det(S_1 - fS_2) = 0$ and $0 < \lambda_1 \leqslant \lambda_2 \leqslant \ldots \leqslant \lambda_m < \infty$ be the eigenvalues of $\det(\Sigma_1 - \lambda\Sigma_2) = 0$. The criterion W_1 for testing the hypothesis $\delta\Lambda = I_m$, $\delta > 0$ (known) is given by

$$W_1 = \prod_{i=1}^{m} (1 - e_i) = \det(I_m - E_1) \tag{5.4.19}$$

where I_m is an identity matrix of order m, $\Lambda = \operatorname{diag}(\lambda_1, \ldots, \lambda_m)$ $E_1 = \operatorname{diag}(e_1, \ldots, e_m)$, $e_i = \delta f_i/(1 + \delta f_i)$, $i = 1, \ldots, m$. The hth moment of W_1 is derived by Pillai, Al-Ami and Jouris (1969) as,

$$E(W_1^h) = \frac{\Gamma_m\left(\frac{n}{2}\right)\Gamma_m\left(\frac{n_2}{2}+h\right)}{\Gamma_m\left(\frac{n_2}{2}\right)\Gamma_m\left(\frac{n}{2}+h\right)} (\det \delta\Lambda)^{-n_1/2} \, {}_2F_1\left(\frac{n}{2}, \frac{n_1}{2}; \frac{n}{2}+h; I_m - (\delta\Lambda)^{-1}\right), \; n = n_1 + n_2. \tag{5.4.20}$$

Thus the density of W_1 is available from the inverse Mellin transform. That is,

$$g_1(w_1) = \frac{\Gamma_m\left(\frac{n}{2}\right)(\det \delta\Lambda)^{-n_1/2}}{\Gamma_m\left(\frac{n_2}{2}\right)} \sum_{k=0}^{\infty}\sum_{K} \frac{\left(\frac{n}{2}\right)_K \left(\frac{n_1}{2}\right)_K}{k!} C_K(I_m - (\delta\Lambda)^{-1})$$
$$\times w_1^{n_2/2-1} G_{m,m}^{m,0}\left[w_1 \middle| \begin{matrix} \frac{n_1}{2} + k_i - \frac{i-1}{2}, i = 1, \ldots, m \\ -\frac{(i-1)}{2}, i = 1, \ldots, m \end{matrix}\right], 0 < w_1 < 1. \tag{5.4.21}$$

Consider the problem of testing hypotheses on regression coefficients. Wilks' criterion in this case is

$$W_2 = \prod_{i=1}^{m} (1 - g_i) \tag{5.4.22}$$

where $g_1, \ldots, g_m$ are the eigenvalues given by $\det(S_1 - g(S_1 + S_2)) = 0$ where the $m \times m$ matrix S_1 has a noncentral Wishart distribution on s degrees of freedom with noncentrality parameter Ω and S_2 has a central

Wishart distribution with t degrees of freedom. Then the hth moment of W_2 can be seen to be

$$E(W_2^h) = \frac{\Gamma_m\left(h+\frac{t}{2}\right)\Gamma_m\left(\frac{n}{2}\right)}{\Gamma_m\left(\frac{t}{2}\right)\Gamma_m\left(h+\frac{n}{2}\right)}\,{}_1F_1\left(h; h+\frac{n}{2}; -\Omega\right),\ n = s+t \tag{5.4.23}$$

Hence the density of W_2 is given by

$$g_2(w_2) = \frac{\Gamma_m\left(\frac{n}{2}\right)}{\Gamma_m\left(\frac{t}{2}\right)} e^{-\mathrm{tr}\,\Omega} \sum_{k=0}^{\infty}\sum_{K} \frac{\left(\frac{n}{2}\right)_K C_K(\Omega)}{k!} \tag{5.4.24}$$

$$\times\, w_2^{t/2-1}\, G_{m,m}^{m,0}\left(w_2 \,\middle|\, \begin{matrix} \frac{s}{2}+k_i-\frac{i-1}{2},\ i=1,\ldots,m \\ -\frac{(i-1)}{2},\ i=1,\ldots,m \end{matrix}\right),\ 0 < w_2 < 1.$$

REMARK. In (5.4.21) and (5.4.24) computable representations are available by using the technique given in Mathai and Saxena (1973a) or Mathai (1972b).

Before concluding this section it should be pointed out that the differential equations are available only in the case of ${}_0F_0$, ${}_0F_1$, ${}_1F_0$, ${}_1F_1$, ${}_2F_1$, ${}_2F_2$ and ${}_3F_2$ in the case of functions of matrix arguments. The differential equations satisfied by a general ${}_pF_q(\cdot)$ or a G-function are still open problems when the arguments are matrices. The differential equation satisfied by an H-function, even when the argument is scalar, is still an open problem.

5.5 GENERALIZED PROBABILITY LAWS

When dealing with probability models to describe experimental situations or natural phenomena, it is often seen that a general probability model with a number of parameters at the choice of the experimenter is a better fit to the experimental situation than a particular model with only one parameter at the choice of the experimenter. Several authors have looked into the problem of studying the common properties of families of probability laws rather than studying these properties for each and every individual member of the family. This topic is known as the topic of generalized probability laws.

In the univariate case very general classes were considered by Mathai and Saxena (1969, 1972). Extensions of some of these results to matrix variable cases were considered by Roux (1975). For the sake of illustration one such probability law is given here. A few more are given in the exercises at the end of this chapter.

Let X and R be $m \times m$ matrices. Consider the following functions.

$$f(X) = C\,(\det X)^{\alpha-(m+1)/2}\,(\det(I-X))^{\beta-(m+1)/2}$$

$$_pF_q\,(a_1,\ldots,a_p;\, b_1,\ldots,b_q;\, RX) \qquad (5.5.1)$$

for

$$0 < X < I,\ \alpha > (m+1)/2 - 1,\ \beta > (m+1)/2 - 1,\ q \geqslant p$$

$$\text{or} \qquad p = q+1 \text{ and } 0 < RX < I$$

where C is a normalizing constant given by,

$$C^{-1} = [\Gamma_m(\alpha+\beta)]^{-1}\,\Gamma_m(\alpha)\,\Gamma_m(\beta)\ {}_{p+1}F_{q+1}(a_1,\ldots,a_p,\alpha;\, b_1,\ldots,b_q,\alpha+\beta;\, R).$$

For convenience we shall assume that the matrix random variable X is not complex. If R is a null matrix then (5.5.1) gives

$$f(X) = \frac{\Gamma_m(\alpha+\beta)}{\Gamma_m(\alpha)\,\Gamma_m(\beta)}\,(\det X)^{\alpha-(m+1)/2}\,(\det(I-X))^{\beta-(m+1)/2},\ 0 < X < I \qquad (5.5.2)$$

This is the multivariate beta distribution in the central case. By specializing the parameters and R in (5.5.1) we get the central and noncentral cases of a large number of probability laws for the matrix random variable X because a large number of elementary functions are special cases of a $_pF_q(\cdot)$. In a similar fashion, by incorporating a hypergeometric function of matrix argument to a gamma type distribution we can get a wide class of densities which include the central and noncentral Wishart distributions. This can be easily noticed from (5.4.4) and (5.4.7).

Other statistical problems condsidered in Mathai and Saxena (1973a chapter 6) can also be given generalizations to matrix variable cases. Since the generalizations do not seem to present any particular difficulties they will not be considered here.

EXERCISES

Prove the following results for functions of matrix argument. All the matrices appearing in these problems are $m \times m$ symmetric positive definite matrices unless otherwise stated.

5.1 $$\int_{\Lambda>0} e^{-\operatorname{tr}\Lambda Z}\ {}_1F_1(\alpha;\beta;\, U\Lambda)\,(\det\Lambda)^{\beta-(m+1)/2}\, d\Lambda$$

$$= \Gamma_m(\beta)\,(\det Z)^{-\beta}\det(I-UZ^{-1})^{-\alpha},\ R(Z) > 0,\ R(Z) > R(M),$$

$$R(\beta) > \frac{m+1}{2} - 1.$$

5.2 $$\int_0^B (\det Z)^{\rho-(m+1)/2} (\det (I + AZ))^{-\nu} dZ$$

$$= \frac{\Gamma_m(\rho)\,\Gamma_m\left(\frac{m+1}{2}\right)}{\Gamma_m\left(\rho + \frac{k+1}{2}\right)} \det (B)^{\rho} \; {}_2F_1\left(\rho, \nu; \rho + \frac{k+1}{2}; -AB\right),$$

$$R(\rho) > \frac{m+1}{2} - 1.$$

5.3 $$\int_{Z>0} (\det (I + Z))^{\nu} (\det (I + AZ))^{\mu} (\det Z)^{\rho-1} \, dZ$$

$$= \frac{\Gamma_m(\rho)\,\Gamma_m(-\mu-\nu-\rho)}{\Gamma_m(-\mu-\nu)} \; {}_2F_1(\mu, \rho; -\mu-\nu; I - A)$$

for $$-R(\mu+\nu) > R(\rho) > \frac{m+1}{2} - 1.$$

5.4 $$\int_0^I (\det Z)^{\rho-(m+1)/2} (\det (I - Z))^{\nu} (\det (I + AZ))^{\mu} \, dZ$$

$$= \frac{\Gamma_m\left(\nu + \frac{m+1}{2}\right)\Gamma_m(\rho)}{\Gamma_m\left(\nu + \rho + \frac{m+1}{2}\right)} \; {}_2F_1\left(-\mu, \rho; \nu + \rho + \frac{m+1}{2}; -A\right)$$

for $$R(\rho) > \frac{m+1}{2} - 1, \; R(\nu) > -1.$$

5.5 $$\int_0^A (\det Z)^{\rho-(m+1)/2} e^{-\operatorname{tr} BZ} \, dZ = (\det A)^{s} \frac{\Gamma_m(\rho)\,\Gamma_m\left(\frac{m+1}{2}\right)}{\Gamma_m\left(\rho + \frac{m+1}{2}\right)}$$

$$\times \, {}_1F_1\left(\rho; \rho + \frac{m+1}{2}; -A^{1/2} B A^{1/2}\right) \quad \text{for} \quad R(\rho) > \frac{m+1}{2} - 1.$$

5.6 $$\int_{Z>0} (\det Z)^{\rho-(m+1)/2} e^{-\operatorname{tr} BZ} \, \gamma(a, Z) \, dZ$$

$$= \frac{\Gamma_m(a)\,\Gamma_m\left(\frac{m+1}{2}\right)\Gamma_m(\rho + a)}{\Gamma_m\left(a + \frac{m+1}{2}\right)} (\det B)^{-(\rho+a)}$$

$${}_2F_1\left(a, a+\rho;\ a+\frac{m+1}{2};\ -B^{-1}\right), \text{ for } R(B) > 0,\ R(a+\rho) > \frac{m+1}{2} - 1,$$

where $\gamma(a, Z)$ is the generalized incomplete gamma function.

5.7 $$G_{0,1}^{1,0}(Z \mid a) = (\det Z)^a e^{-\operatorname{tr} Z}$$

5.8 $$G_{1,1}^{1,1}\left(Z \,\middle|\, \begin{matrix}(m+1)/2-\rho \\ a\end{matrix}\right) = \Gamma_m(\rho+a)(\det Z)^a (\det(I+Z))^{-(\rho+a)}.$$

5.9 $$\int_{\Lambda>0} (\det \Lambda)^{\rho-(m+1)/2} e^{-\operatorname{tr}\Lambda} G_{0,1}^{1,0}(\Lambda Z \mid a)\, d\Lambda = G_{1,1}^{1,1}\left(Z \,\middle|\, \begin{matrix}(m+1)/2-\rho \\ a\end{matrix}\right).$$

5.10 $$\int_0^I (\det Z)^{a-(m+1)/2} (\det(I-Z))^{b-(m+1)/2}\, {}_2F_1(\alpha, \beta; \gamma; \Lambda Z)\, dZ$$

$$= \frac{\Gamma_m(a), \Gamma_m(b)}{\Gamma_m(a+b)}\, {}_3F_2(a, \alpha, \beta; a+b, \gamma; \Lambda)$$

for

$$R(a), R(b) > \frac{m+1}{2} - 1.$$

5.11 Obtain Euler's formula for ${}_2F_1(\cdot)$ from Problem 5.10.

5.12 $${}_2F_1(a, b; c; I) = \frac{\Gamma_m(c)\,\Gamma_m(c-a-b)}{\Gamma_m(c-a)\,\Gamma_m(c-b)},$$

$$R(c), R(c-a-b) > \frac{m+1}{2} - 1.$$

5.13 $$\int_0^I (\det Z)^{\gamma-(m+1)/2} (\det(I-Z))^{b-(m+1)/2}\, {}_2F_1(\alpha, \beta; \gamma; Z)\, dZ$$

$$= \frac{\Gamma_m(\gamma)\,\Gamma_m(\gamma+b-\alpha-\beta)\,\Gamma_m(b)}{\Gamma_m(\gamma+b-\alpha)\,\Gamma_m(\gamma+b-\beta)}$$

for

$$R(\gamma), R(b) > \frac{m+1}{2} - 1,\ R(\gamma+b-\alpha-\beta) > 0.$$

5.14 $$\int_0^I (\det Z)^{a-(m+1)/2} (\det(I-Z))^{b-(m+1)/2}\, {}_2F_1(\alpha, a+b; \gamma; Z)\, dZ$$

$$= \frac{\Gamma_m(a)\,\Gamma_m(b)\,\Gamma_m(\gamma)\,\Gamma_m(\gamma-a-\alpha)}{\Gamma_m(a+b)\,\Gamma_m(\gamma-\alpha)\,\Gamma_m(\gamma-a)}$$

for

$$R(a), R(b) > \frac{m+1}{2} - 1,\ R(\gamma-a-\alpha) > 0.$$

5.15 $$\int_0^{\Omega} (\det Z)^{a-(m+1)/2}\, G^{p,q}_{r,s}\left[Z\Lambda \,\middle|\, \begin{matrix} a_1, \ldots, a_r \\ b_1, \ldots, b_s \end{matrix} \right] dZ$$

$$= (\det \Omega)^a\, \Gamma_m\left(\frac{m+1}{2}\right) G^{p,q+1}_{r+1,s+1}\left[\Omega\Lambda \,\middle|\, \begin{matrix} \frac{m+1}{2} - a, a_1, \ldots, a_r \\ b_1, \ldots, b_s, -a \end{matrix} \right]$$

for

$$R(a + \min b_j) > \frac{m+1}{2} - 1,\ j = 1, \ldots, p.$$

5.16 From Problem 5.15 deduce,

$$\int_0^{\Omega} (\det Z)^{a-(m+1)/2}\, {}_rF_s(a_1, \ldots, a_r;\ b_1, \ldots, b_s;\ Z\Lambda)\, dZ$$

$$= (\det \Omega)^a \frac{\Gamma_m(a)\, \Gamma_m\left(\frac{m+1}{2}\right)}{\Gamma_m\left(a + \frac{m+1}{2}\right)}\, {}_{r+1}F_{s+1}\left(a, a_1, \ldots, a_r;\ a + \frac{m+1}{2},\right.$$

$$\left. b_1, \ldots, b_s;\ \Lambda\Omega\right), \text{ for } R(a) > \frac{m+1}{2} - 1.$$

5.17 From Problem 5.16 deduce the incomplete gamma function and the incomplete beta function.

5.18 $$\frac{2^{m(m+1)/2}}{(2\pi i)^{m(m+1)/2}} \int_{R(Z)>0} e^{\operatorname{tr} SZ} (\det Z)^{-\rho}\, C_K(Z^{-1})\, dZ$$

$$= \frac{1}{\Gamma_m(\rho, K)} (\det S)^{\rho-(m+1)/2}\, C_K(S)$$

where $C_K(S)$ is the zonal polynomial.

5.19 $$\int_0^{\Omega} (\det Z)^{\rho-(m+1)/2} (\det(\Omega - Z))^{\nu-(m+1)/2}\, C_K(Z)\, dZ$$

$$= F(\Omega)\, (\det \Omega)^{\rho+\nu-(m+1)/2} \text{ where } F(\Omega) = C_K(\Omega)\, \frac{F(I)}{C_K(I)}.$$

5.20 Write down the values of the integrals in Problems 5.1 to 5.16 in series involving zonal polynomials.

5.21 $${}_0F_1\left(\frac{n}{2}; \frac{1}{4} XX'\right) = \int_{0(n)} e^{\operatorname{tr} XH}\, dH.$$

5.22 $$\int_{0(m)} (\det(I + SHTH'))^{-a}\, (dH)$$

$$= {}_1F_0(a;\ -S, T) \quad \text{for } S > 0,\ T > 0$$

where the hypergeometric function of two arguments is defined as follows:

$$ {}_pF_q(a_1,\ldots a_p;\ b_1,\ldots,b_q;\ S,T)=\sum_{k=0}^{\infty}\sum_{K}\frac{(a_1)_K\ldots(a_p)_K\ C_K(S)\ C_K(T)}{(b_1)_K\ldots(b_q)_K\ C_K(I_m)\ k!} $$

5.23 Show that

$$ \frac{1}{2^m}\int_{0(m)} {}_0F_1\left(\frac{s}{2};\ \Omega\, G^{1/2}\ HRH'G^{1/2}\right) d(H) $$

$$ =\frac{\pi^{m^2/2}}{\Gamma_m\left(\frac{m}{2}\right)}\sum_{k=0}^{\infty}\sum_{K}\frac{C_K(\Omega\ G)\ \ C_K(R)}{\left(\frac{s}{2}\right)_K\ \ C_K(I)\ \ \ k!}. $$

5.24 If the $m\times m$ matrix A has the noncentral Wishart distribution on s degrees of freedom and noncentrality parameter Ω and B has the Wishart distribution with t degrees of freedom, the covariance matrix in each case being Σ, then show that the probability density function of the roots of the matrix $R=A(A+B)^{-1}$ is given by

$$ \frac{\pi^{m^2/2}\,\Gamma_m\left(\frac{s+t}{2}\right)}{\Gamma_m\left(\frac{s}{2}\right)\Gamma_m\left(\frac{t}{2}\right)\Gamma_m\left(\frac{m}{2}\right)}\, e^{-\operatorname{tr}\Omega}\ \left(\Pi\, r_i\right)^{(s-m-1)/2}\ \Pi\,(1-r_i)^{(t-m-1)/2} $$

$$ \times\prod_{i<j}(r_i-r_j)\ \sum_{k=0}^{\infty}\sum_{K}\frac{\left(\frac{s+1}{2}\right)_K\ C_K(\Omega)\ C_K(R)}{\left(\frac{s}{2}\right)_K\ C_K(I)\ k!} $$

where r_i's are the latent roots of the matrix $A(A+B)^{-1}$ (Constantine, 1963).

Hint. Transform $A\to\frac{1}{2}\,\Sigma^{-1/2}\,A\,\Sigma^{-1/2}$, $B\to\frac{1}{2}\,\Sigma^{-1/2}\,B\,\Sigma^{-1/2}$, obtain the joint density of A and B then transform $G=A+B$, $R=G^{-1/2}\ AG^{-1/2}$ $\Rightarrow G>0,\ 0<R<I.$ Diagonalize R by an orthogonal matrix H, that is H^1RH is diagonal $\Rightarrow dR=\prod_{i<j}(r_i-r_j)\ \Pi\, dr_i\ d(H)$ and integrate over $d(H)$.

5.15 Show that

$$ E(W^h)=\frac{\Gamma_m\left(h+\frac{t}{2}\right)\Gamma_m\left(\frac{s+t}{2}\right)}{\Gamma_m\left(\frac{t}{2}\right)\Gamma_m\left(h+\frac{s+t}{2}\right)}\ {}_1F_1\left(h;\ h+\frac{s+t}{2};\ -\Omega\right) $$

where $W=\Pi\,(1-r_i)=\det(I-R)$, R and r_i are defined in Problem 5.24.

5.26 Let the columns of $\begin{pmatrix}X_1\\X_2\end{pmatrix}$ be independent normal $p+q-$variates with mean vector null and covariance matrix $\Sigma = \begin{bmatrix} \Sigma_{11} & \Sigma_{12} \\ \Sigma_{21} & \Sigma_{22} \end{bmatrix}$, $\Sigma_{21} = \Sigma_{12}'$, $p < q$, $p + q \leqslant n$, n being the sample size. Let $R^2 = \text{diag}(r_1^2, \ldots, r_p^2)$ where r_i^2 are the eigenvalues of,

$$\det [X_1 X_1' (X_2 X_2')^{-1} X_2 X_1' - r^2 X_1 X_1'] = 0$$

and let $P^2 = \text{diag}(\rho_1^2, \ldots, \rho_p^2)$ where ρ_i^2 are the eigenvalues of

$$\det [\Sigma_{12} \Sigma_{22}^{-1} \Sigma_{12}' - \rho^2 \Sigma_{11}] = 0. \quad \text{Let } W_3 = \prod_{i=1}^{p} (1 - r_i^2)$$

Show that

$$E(W_3^h) = \frac{\Gamma_p\left(\frac{n}{2}\right) \Gamma_p\left(\frac{n-q}{2} + h\right)}{\Gamma_p\left(\frac{n-q}{2}\right) \Gamma_p\left(\frac{n}{2} + h\right)} \det (I_p - P^2)^{n/2}$$
$$\times \, {}_2F_1\left(\frac{n}{2}, \frac{n}{2}; \frac{n}{2} + h; P^2\right).$$

5.27 Evaluate the density of W_3 in Problem 5.26 and represent it in a computable form.

5.28 Work out the nonnull distributions corresponding to the various null distributions discussed in Chapter 6 of Mathai and Saxena (1973a). [A number of these are open problems.]

5.29 Show that the following is a probability law. Obtain the central and noncentral Wishart distributions as particular cases of the following probability law.

$$f(X) = C (\det X)^{\beta - (m+1)/2} e^{-\text{tr } BX} \, {}_pF_q (a_1, \ldots, a_p; b_1, \ldots, b_q; BRBX), \; X > 0$$

where C is a normalizing constant.

5.30 By using the property that the total probability is unity show that C in Problem 5.29 is given by

$$C^{-1} = (\det B)^{-\beta} \, \Gamma_m(\beta) \, {}_{p+1}F_q (a_1, \ldots, a_p, \beta; b_1, \ldots, b_q; BR)$$

where R and B are $m \times m$ positive definite matrices.

5.31 For what values of C and the parameters the following is a probability law

$$g(Y) = C (\det (I + Y))^{-(\alpha+\beta)} (\det Y)^{\alpha - (m+1)/2}$$
$$\times \, {}_pF_q (a_1, \ldots, a_p; b_1, \ldots, b_q; R(I + Y)^{-1}),$$

for $Y > 0$ where R and Y are $m \times m$ positive definite matrices.

5.32 Evaluate C if the following is a probability law

$$f(X_1, \ldots, X_k) = C \prod_{j=1}^{k} (\det (X_j))^{\alpha_j-(m+1)/2} \det\left(I - \sum_{j=1}^{k} X_j\right)^{\beta-(m+1)/2}$$

$$\times {}_pF_q\left(a_1, \ldots, a_p; b_1, \ldots, b_q; R \sum_{j=1}^{k} X_j\right),\ I - \sum_{j=1}^{k} X_j > 0, 0 < X_j < I.$$

This is the generalization of the Dirichlet distribution to the matrix variable case. All the matrices are $m \times m$ symmetric matrices.

5.33 Show that the normalizing constant C in the following density

$$g(Y_1, \ldots, Y_k) = C \prod_{j=1}^{k} (\det Y_j)^{\alpha_j-(m+1)/2} \left(\det\left(I + \sum_{j=1}^{k} Y_j\right)\right)^{-(\alpha+\beta)}$$

$$\times {}_pF_q\left(a_1, \ldots, a_p; b_1, \ldots, b_q; R\left(I + \sum_{j=1}^{k} Y_j\right)^{-1}\right),\ Y_j > 0, j = 1, \ldots, k$$

is given by

$$C^{-1} = [\Gamma_m(\alpha+\beta)]^{-1} \prod_{j=1}^{k} \Gamma_m(\alpha_j)\, \Gamma_m(\beta)$$

$$\times {}_{p+1}F_{q+1}(a_1, \ldots, a_p, \alpha; b_1, \ldots, b_q, \alpha+\beta; R)$$

where

$$\alpha = \sum_{j=1}^{k} \alpha_j.$$

5.34 Evaluate the following expected values. $E(\det X)^\gamma$, $E(\det Y)^\gamma$, $E[\det(X_1)^{\gamma_1}\det(X_2)^{\gamma_2}\ldots\det(X_k)^{\gamma_k}]$, $E[\det(Y_1)^{\gamma_1}\ldots\det(Y_k)^{\gamma_k}]$ in Problems 5.29, 5.31, 5.32 and 5.33 respectively.

5.35 Evaluate the integral

$$\int_{S>0} e^{-\operatorname{tr}\Sigma^{-1}S/2} (\det S)^{(\alpha+\beta)/2-(m+1)/2}\, {}_0F_1\left(\frac{\beta}{2}; \frac{1}{2}\Omega\Sigma^{-1}S^{1/2}LS^{1/2}\right) dS$$

for $\quad S > 0,\ \Omega > 0,\ \Sigma > 0,\ 0 < L < I.$

Appendix

Here we give the definitions of nearly all the elementary special functions of one variable and important functions of two or more variables. A detailed description of the various properties of these functions can be found in the monograph by Magnus, Oberhettinger and Soni (1966), Erdélyi, A. et al. (1953, 1955), Rainville (1966) and Kampé de Fériet (1926). These definitions can be used in deriving the relations that exist between the elementary special functions and the *H*-function. On account of the general character of the *H*-function results for elementary special functions can be derived as special cases. Hence we have also presented the *H*-function in terms of elementary special functions and vice versa, which may be found useful by the readers in deriving the results for the elementary special functions from the formulae of the *H*-function.

A1 GAMMA FUNCTIONS AND RELATED FUNCTIONS

Gamma Function

$$\Gamma(z) = \int_0^\infty e^{-x} x^{z-1}\, dx, \quad R(z) > 0$$

$$\frac{1}{\Gamma(z)} = \frac{1}{2\pi i} \int_{c-i\infty}^{c+i\infty} e^t t^{-z}\, dt, \; c > 0 \text{ and } R(z) > 0.$$

The modified forms of the Gauss and Legendre's multiplication theorem for the gamma function are as follows.

$$(a)_{nk} = k^{nk} \prod_{s=1}^{k} \left\{ \left(\frac{a+s-1}{k} \right) \right\}^n$$

$$(a)_{-nk} = \left(-\frac{1}{k}\right)^{nk} \frac{1}{\prod_{s=1}^{k} \left\{\left(\frac{s-a}{k}\right)\right\}_n}$$

where the Pochammer's symbol

$$(a)_r = \frac{\Gamma(a+r)}{\Gamma(a)} = a(a+1)\ldots(a+r-1);\ (a)_0 = 1.$$

PSI FUNCTION

$$\psi(z) = \frac{d}{dz}\log\Gamma(z) = \frac{\Gamma'(z)}{\Gamma(z)} = \int_0^\infty [t^{-1}e^{-t} - (1-e^{-t})^{-1}e^{-tz}]\,dt,$$

$$R(z) > 0.$$

$$= -\gamma + (z-1)\sum_{k=0}^{\infty}[(k+1)(z+k)]^{-1}.$$

$$\gamma \approx 0.5772156649\ldots.$$

EULER'S DILOGARITHM

$$L_2(z) = \sum_{n=1}^{\infty}\frac{z^n}{n^2} = -\int_0^z \frac{\log(1-z)}{z}\,dz.$$

BETA FUNCTION

$$B(x, y) = \frac{\Gamma(x)\,\Gamma(y)}{\Gamma(x+y)} = \int_0^1 t^{x-1}(1-t)^{y-1}\,dt.$$

$$R(x) > 0 \text{ and } R(y) > 0.$$

RIEMANN'S ZETA FUNCTION AND RELATED FUNCTIONS

$$\zeta(s) = \sum_{n=1}^{\infty} n^{-s}$$

$$\zeta(t) = -\frac{1}{2}\left(t^2 + \frac{1}{4}\right)\pi^{-it/2-1/4}\,\Gamma\left(\frac{it}{2} + \frac{1}{4}\right)\zeta\left(it + \frac{1}{2}\right).$$

$$\zeta(z, a) = \sum_{n=0}^{\infty}(n+a)^{-z}.$$

$$\phi(z, s, \gamma) = \sum_{n=0}^{\infty}\frac{z^n}{(\gamma+n)^s}.$$

ORTHOGONAL POLYNOMIALS

BESSEL POLYNOMIAL

$$Y_n(x, a, b) = \sum_{r=0}^{n}\frac{(-n)_r(a+n-1)_r\left(-\frac{x}{b}\right)^r}{r!}$$

$$= {}_2F_0\left(-n, a+n-1; -; -\frac{x}{b}\right)$$

ORTHOGONALITY PROPERTY

$$\int_0^\infty x^{1-a} e^{-x} Y_n(1, a, x)\, Y_m(1, b, x)\, dx$$

$$= \begin{cases} 0 & \text{if } m \neq n \\ n! \quad \Gamma(2-a-n) & \text{if } m = n \end{cases}.$$

GEGENBAUER POLYNOMIAL

$$C_n^\nu(x) = \frac{(-2)^n (\nu)_n}{n!\,(2\nu+n)_n} (1-x^2)^{1/2-\nu} \frac{d^n}{dx^n} (1-x^2)^{\nu+n-1/2}.$$

Orthogonality property

$$\int_{-1}^{1} (1-x^2)^{\nu-1/2}\, C_m^\nu(x)\, C_n^\nu(x)\, dx$$

$$= \frac{\pi\, 2^{1-2\nu}\, \Gamma(n+2\nu)}{n!\,(n+\nu)\{\Gamma(\nu)\}^2} \delta_{mn},\ R(\nu) > -\frac{1}{2},$$

where

$$\delta_{mn} = \begin{cases} 0, & \text{if} \quad m \neq n \\ 1, & \text{if} \quad m = n. \end{cases}$$

LEGENDRE POLYNOMIAL

$$P_n(x) = \frac{1}{2^n (n)!} \frac{d^n}{dx^n} (x^2-1)^n.$$

Orthogonality property

$$\int_{-1}^{1} P_m(x)\, P_n(x)\, dx = \frac{2}{2n+1} \delta_{mn}.$$

TCHEBICHEF POLYNOMIALS

$$T_n(x) = \cos(n \cos^{-1} x)$$

$$U_n(x) = \frac{[\sin(n+1)\cos^{-1} x]}{\sin(\cos^{-1} x)}.$$

Orthogonality property

$$\int_{-1}^{1} (1-x^2)^{-1/2}\, T_m(x)\, T_n(x)\, dx = \frac{\pi}{2} \delta_{mn}$$

$$\int_{-1}^{1} (1 - x^2)^{-1/2}\, T_m(x)\, T_n(x)\, dx = \frac{\pi}{2}\,\delta_{mn}.$$

JACOBI POLYNOMIAL

$$P_n^{(\alpha,\beta)}(x) = \frac{(-1)^n}{2^n n!}(1 - x)^{-\alpha}(1 + x)^{-\beta}\frac{d^n}{dx^n}[(1 - x)^{n+\alpha}(1 + x)^{n+\beta}]$$

Orthogonality property

$$\int_{-1}^{1} (1 - x)^{\alpha}(1 + x)^{\beta} P_m^{(\alpha,\beta)}(x)\, P_n^{(\alpha,\beta)}(x)\, dx$$

$$= \frac{2^{\alpha+\beta+1}}{\alpha + \beta + 2n + 1}\cdot\frac{\Gamma(\alpha + n + 1)\,\Gamma(\beta + n + 1)}{(n)!\,\Gamma(\alpha + \beta + n + 1)}\,\delta_{mn},$$

where

$$R(\alpha) > -1,\; R(\beta) > -1.$$

LAGUERRE POLYNOMIAL

$$L_n^{(\alpha)}(x) = \frac{e^x x^{-\alpha}}{n!}\frac{d^n}{dx^n}(e^{-x}\, x^{n+\alpha}).$$

$$L_n^0(x) = L_n(x).$$

Orthogonality property

$$\int_0^{\infty} e^{-x} x^{\alpha} L_m^{(\alpha)}(x)\, L_n^{(\alpha)}(x)\, dx = \frac{\Gamma(\alpha + n + 1)}{n!}\,\delta_{mn},\; R(\alpha + n + 1) > 0.$$

CHARLIER POLYNOMIAL

$$P_n(x; a) = n!\, \bar{a}^n\, L_n^{x-n}(a).$$

A2 HYPERGEOMETRIC FUNCTIONS

GAUSS' HYPERGEOMETRIC FUNCTION

$${}_2F_1(\alpha, \beta; \gamma; z) = \sum_{r=0}^{\infty} \frac{(\alpha)_r\,(\beta)_r}{(\gamma)_r\, r!} z^r,$$

where γ is neither zero nor a negative integer,

$|z| < 1$; $z = 1$ and $R(\gamma - \alpha - \beta) > 0$; $z = -1$, and $R(\gamma - \alpha - \beta) > -1$.

$$= \frac{\Gamma(\gamma)}{\Gamma(\alpha)\,\Gamma(\beta)}\frac{1}{2\pi i}\int_{-i\infty}^{+i\infty} \frac{\Gamma(\alpha + s)\,\Gamma(\beta + s)}{\Gamma(\gamma + s)}\,\Gamma(-s)(-z)^s\, ds$$

where $|\arg(-z)| < \pi$ and the path of integration is indented, if necessary, in such a manner so as to separate the poles at $s = 0, 1, 2, \ldots$ from the poles at $s = -\alpha - n$, $s = -\beta - n$ $(n = 0, 1, 2, \ldots)$ of the integrand.

$$\frac{\Gamma(\alpha)\,\Gamma(\beta)}{\Gamma(\gamma)}\ {}_2F_1(\alpha, \beta; \gamma; x + y)$$

$$= -\frac{1}{4\pi^2}\int_{-i\infty}^{+i\infty}\int_{-i\infty}^{+i\infty}\frac{\Gamma(-s)\,\Gamma(-t)\,\Gamma(\alpha+s+t)\,\Gamma(\beta+s+t)\,(-x)^s\,(-y)^t}{\Gamma(\gamma+s+t)}\,ds\,dt$$

GENERALIZED HYPERGEOMETRIC FUNCTION

$${}_pF_q(\alpha_1, \ldots, \alpha_p; \beta_1, \ldots, \beta_q; z) = \sum_{r=0}^{\infty}\frac{\prod_{j=1}^{p}(\alpha_j)_r\, z^r}{\prod_{j=1}^{q}(\beta_j)_r\,(r)!}$$

$$(p \leqslant q;\ p = q + 1 \text{ and } |z| < 1)$$

$$= \frac{\prod_{j=1}^{p}\Gamma(\beta_j)}{\prod_{=1}^{q}\Gamma(\alpha_j)}$$

$$\times \frac{1}{2\pi i}\int_{-i\infty}^{+i\infty}\frac{\Gamma(\alpha_1 + s)\ldots(\alpha_p + s)\,\Gamma(-s)\,(-z)^s\,ds}{\Gamma(\beta_1 + s)\,\ldots\,\Gamma(\beta_q + s)}$$

where $|\arg(-z)| < \pi$ and the path of integration is indented, if necessary, in such a manner that the poles of $\Gamma(-s)$ are separated from those of $\Gamma(a_j + s)$, $(j = 1, \ldots, p)$.

$$S_n(b_1, b_2, b_3, b_4; z) = \sum_{r=1}^{n}\frac{\prod_{j=1}^{n}{}'\,\Gamma(b_j - b_r)}{\prod_{j=n+1}^{4}\Gamma(1 + b_r - b_j)}\,z^{1+2b_r}$$

$$\times\ {}_0F_3(1 + b_r - b_1, \ldots * \ldots, 1 + b_r - b_4; (-1)^n z^2),$$

where the prime in Π and the asterisk in ${}_0F_3$ indicate that the term containing $b_h - b_h$ is to be omitted.

An empty product is always interpreted as unity.

INCOMPLETE BETA FUNCTION

$$\text{(i)}\quad B_x(p, q) = \int_0^x t^{p-1}(1 - t)^{q-1}\,dt$$

$$= p^{-1}\,x^p\ {}_2F_1(p, 1 - q; p + 1; x).$$

(ii) $I_x(p,q) = \dfrac{B_x(p,q)}{B_1(p,q)}$.

Legendre Functions

$$P_\nu^\mu(z) = \frac{1}{\Gamma(1-\mu)}\left(\frac{z+1}{z-1}\right)^{\mu/2} {}_2F_1\left(-\nu, \nu+1; 1-\mu; \frac{1-z}{2}\right).$$

$$Q_\nu^\mu(z) = \frac{e^{\mu\pi i}\,\Gamma(\tfrac{1}{2})\,\Gamma(\mu+\nu+1)}{2^{\nu+1}\,\Gamma(\nu+\tfrac{3}{2})} z^{-\mu-\nu-1}(z^2-1)^{\mu/2}$$

$$\times {}_2F_1\left(\frac{\mu+\nu+1}{2}, \frac{\mu+\nu+2}{2}; \nu+\frac{3}{2}; \frac{1}{z^2}\right).$$

z in the complex plane cut along the real axis from -1 to $+1$.

$$P_\nu^\mu(x) = \frac{1}{\Gamma(1-\mu)}\left(\frac{1+x}{1-x}\right)^{\mu/2} {}_2F_1\left(-\nu, \nu+1; 1-\mu; \frac{1-x}{2}\right), \quad -1 < x < 1.$$

$$Q_\nu^\mu(x) = \frac{e^{-i\mu\pi}}{2}\left[e^{-\mu\pi i/2}\,Q_\nu^\mu(x+i0) + e^{\mu\pi i/2}\,Q_\nu^\mu(x-i0)\right], \quad -1 < x < 1.$$

$$P_\nu^\mu(z) = P_\nu^0(z), \quad Q_\nu(x) = Q_\nu^0(z).$$

A3 CONFLUENT HYPERGEOMETRIC FUNCTIONS

For a detailed account of these functions see the works of Slater (1960) and Tricomi, F. G. [Functioni Ipergeometrische Confluent Ed. Cremorise] Rome (1954).

Whittaker Functions

$$M_{k,m}(z) = z^{m+1/2}\,e^{z/2}\,{}_1F_1(\tfrac{1}{2}+k+m; 2m+1; -z).$$

$$= z^{m+1/2}\,e^{-z/2}\,{}_1F_1(\tfrac{1}{2}-k+m; 2m+1; z).$$

$$= \frac{\Gamma(1+2m)}{\Gamma(\tfrac{1}{2}-k+m)\,\Gamma(\tfrac{1}{2}+k+m)} e^{-z/2} z^{m+1/2}$$

$$\times \int_0^1 e^{-zt}\, t^{m-k-1/2}(1-t)^{m+k-1/2}\,dt,$$

$$R(\tfrac{1}{2} \pm k + m) > 0, \ |\arg z| < \pi$$

$$= \frac{\Gamma(1+2m)}{\Gamma(\tfrac{1}{2}-k+m)} e^{-z/2} z^{m+1/2}$$

$$\times \int_{c-i\infty}^{c+i\infty} \frac{\Gamma(-s)\,\Gamma(\tfrac{1}{2}-k+m+s)}{\Gamma(1+2m+s)} (-z)^s\,ds$$

$$|\arg z| < \frac{\pi}{2},\ 2m \neq 0, -1, -2, \ldots$$

$$W_{k,m}(z) = \sum_{m,-m} \frac{\Gamma(-2m)}{\Gamma(\frac{1}{2} - k - m)} M_{k,m}(z),$$

where the symbol $\sum_{m,-m}$ indicates that to the expression following it a similar expression in which m has been replaced by $-m$ is to be added.

$$W_{k,m}(z) = \frac{2z^{1/2}\, e^{-z/2}}{\Gamma(\frac{1}{2} - k + m)\, \Gamma(\frac{1}{2} - k - m)} \int_0^\infty e^{-x}\, x^{-k-1/2}\, K_{2m}(2(zx)^{1/2})\, dx$$

$R(\frac{1}{2} - k \pm m) > 0$ and $K_\nu(\cdot)$ is the modified Bessel function of the second kind

$$W_{k,m}(z) = \frac{z^k\, e^{-z/2}}{\Gamma(\frac{1}{2} - k + m)\, \Gamma(\frac{1}{2} - k - m)}$$

$$\times \frac{1}{2\pi i} \int_{c-i\infty}^{c+i\infty} \Gamma(-s)\, \Gamma(\tfrac{1}{2} + m - k + s)\, \Gamma(\tfrac{1}{2} - m - k + s)\, z^{-s}\, ds,$$

$$|\arg z| < \frac{3\pi}{2},\ \frac{1}{2} + k \pm m \neq 0, 1, 2, \ldots .$$

PARABOLIC CYLINDER FUNCTION

$$D_\nu(z) = 2^{\nu/2+1/4}\, z^{-1/2}\, W_{\nu/2+1/4,1/4}\left(\frac{z^2}{2}\right)$$

$$= (-1)^n \exp\left(\frac{z^2}{4}\right) \frac{d^n}{dz^n}\, (e^{-z^2/2}).$$

$$D_\nu(z) = 2^{\nu/2-1/4}\, z^{1/2}\, e^{-3z^2/4}$$

$$\times \frac{1}{2\pi i} \int_{-i\infty}^{+i\infty} \frac{\Gamma(s)\, \Gamma\left(\frac{\nu}{2} + \frac{1}{4} - s\right) \Gamma\left(\frac{\nu}{2} - \frac{1}{4} - s\right)}{\Gamma\left(\frac{\nu}{2} + \frac{1}{4}\right) \Gamma\left(\frac{\nu}{2} - \frac{1}{4}\right)} \left(\frac{z^2}{2}\right)^s ds,$$

$$|\arg z| < \frac{3\pi}{4},\ \nu \neq 1, 2, -\frac{1}{2}, -\frac{3}{2}, \ldots .$$

BATEMAN'S FUNCTION

$$k_{2\nu}(z) = \frac{1}{\Gamma(\nu + 1)} W_{\nu,1/2}(2z).$$

THE EXPONENTIAL INTEGRAL AND RELATED FUNCTIONS

$$-E_i(-x) = E_1(x) = \int_x^\infty \frac{e^{-t}\,dt}{t} = \Gamma(0, x), \ -\pi < \arg x < \pi,$$

$$E_i^+(x) = E_i(x + i0), \ E_i^-(x) = E_i(x - i0), \ x > 0.$$

$$\overline{E_i(x)} = \tfrac{1}{2}[E_i^+(x) + E_i^-(x)], \ x > 0.$$

$$l_i(z) = \int_0^\infty \frac{dt}{\log t} = E_i(\log z).$$

$$s_i(x) = -\int_x^\infty \frac{\sin t}{t}\,dt = \frac{1}{2i}[E_i(ix) - E_i(-ix)].$$

$$S_i(x) = \int_0^x \frac{\sin t}{t}\,dt = \frac{\pi}{2} + s_i(x).$$

$$C_i(x) = -\int_x^\infty \frac{\cos t}{t}\,dt = -C_i(x)$$

$$= \tfrac{1}{2}[E_i(ix) + E_i(-ix)].$$

ERROR FUNCTIONS AND RELATED FUNCTIONS

$$\operatorname{Erf}(x) = 2\pi^{-1/2} \int_0^x e^{-t^2}\,dt = \frac{2x}{\sqrt{\pi}}\,{}_1F_1\left(\frac{1}{2}; \frac{3}{2}; -x^2\right).$$

$$\operatorname{Erfc}(x) = 2\pi^{-1/2} \int_0^\infty e^{-t^2}\,dt = 1 - \operatorname{Erf}(x).$$

$$= (\pi x)^{-1/2} e^{-x^2/2} W_{-1/4, 1/4}(x^2).$$

$$C(x) = (2\pi)^{-1/2} \int_0^x t^{-1/2} \cos t\,dt.$$

$$S(x) = (2\pi)^{-1/2} \int_0^x t^{-1/2} \sin t\,dt.$$

INCOMPLETE GAMMA FUNCTIONS

$$\gamma(a, x) = \int_0^x e^{-t}\, t^{a-1}\, dt = a^{-1}\, x^a\, {}_1F_1\,(a;\, a+1;\, -x).$$

$$\Gamma(a, x) = \int_x^\infty e^{-t}\, t^{a-1}\, dt = \Gamma(a) - \gamma(a, x)$$

$$= x^{(a-1)/2}\, e^{-x/2}\, W_{(a-1)/2\,;\,a/2}\,(x).$$

COULOMB WAVE FUNCTIONS

$$F_L(\eta, \sigma) = C_L(\eta)\, \sigma^{L+1}\, e^{-i\sigma}\, {}_1F_1\,(L+1-i\eta;\, 2L+2;\, 2i\sigma),$$

where

$$C_L(\eta) = \frac{2^L\, e^{-\pi\eta/2}\, |\,\Gamma(L+1+i\eta)\,|}{(2L+1)!}.$$

Also see the work of Abramowitz and Stegun [*Handbook of Mathematical Functions* with *Formulas, Graphs and Mathematical Tables.* Appl. Math. Ser. 55, U.S. Govt. Printing Office, Washington, D.C. 1964].

A4 BESSEL FUNCTIONS AND ASSOCIATED FUNCTIONS

$$J_\nu(z) = \frac{\left(\frac{z}{2}\right)^\nu}{\Gamma(\nu+1)}\, {}_0F_1\left(-;\, \nu+1;\, -\frac{z^2}{4}\right)$$

$$= \sum_{r=0}^{\infty} \frac{(-1)^r \left(\frac{z}{2}\right)^{\nu+2r}}{r!\,\Gamma(\nu+r+1)}.$$

$$J_\nu(x) = \frac{1}{4\pi_i} \int_{c-i\infty}^{c+i\infty} \frac{\Gamma\left(\frac{\nu+s}{2}\right)}{\Gamma\left(1+\frac{\nu-s}{2}\right)} \left(\frac{x}{2}\right)^{-s} ds,$$

$$x > 0,\; -R(\nu) < c < 1.$$

$$I_\nu(z) = \sum_{r=0}^{\infty} \frac{\left(\frac{z}{2}\right)^{\nu+2r}}{r!\;\Gamma(\nu+r+1)}$$

$$= \frac{\left(\frac{z}{2}\right)^\nu}{\Gamma(\nu+1)}\, {}_0F_1\left(-\,;\, \nu+1;\, \frac{z^2}{4}\right)$$

$$= e^{-i\nu\pi/2}\, J_\nu\,(ze^{i\pi/2}),\; -\pi < \arg z \leqslant \frac{\pi}{2}.$$

$$I_m\left(\frac{z}{2}\right) = \frac{M_{0,m}(z)}{2^{2m}\,\Gamma(m+1)\,z^{1/2}}.$$

$$Y_\nu(z) = \operatorname{cosec}(\nu\pi)\,[J_\nu(z)\cos(\nu\pi) - J_{-\nu}(z)].$$

$$Y_\nu(x) = \frac{1}{2\pi i}\int_{c-i\infty}^{c+i\infty} \frac{\Gamma\left(s-\frac{\nu}{2}\right)\Gamma\left(s+\frac{\nu}{2}\right)}{\Gamma\left(s-\frac{\nu+1}{2}\right)\Gamma\left(\frac{3+\nu}{2}-s\right)}\left(\frac{x^2}{4}\right)^{-s} ds,$$

$$R(s \pm \nu) > 0.$$

$$K_\nu(z) = \tfrac{1}{2}\sum_{\nu,\,-\nu}\Gamma(-\nu)\,\Gamma(1+\nu)\,I_\nu(z)$$

$$= \frac{\pi}{2\sin\nu\pi}\,[I_{-\nu}(z) - I_\nu(z)]$$

$$K_\nu(z) = \left(\frac{2z}{\pi}\right)^{-1/2} W_{0,\nu}(2z)$$

$$K_\nu(z) = \frac{1}{4\pi i}\int_{c-i\infty}^{c+i\infty}\Gamma\left(s+\frac{\nu}{2}\right)\Gamma\left(s-\frac{\nu}{2}\right)\left(\frac{z^2}{4}\right)^{-s} ds.$$

$$H_\nu^{(1)}(z) = J_\nu(z) + i\,Y_\nu(z)$$

$$H_\nu^{(2)}(z) = J_\nu(z) - i\,Y_\nu(z).$$

$$J_{i_\nu}(z) = \int_\infty^z J_\nu(t)\frac{dt}{t}.$$

$$2\pi^2 H_\nu^{(1)}(z) = -e^{-i\pi i 2}\int_{c-i\infty}^{-c+i\infty}\Gamma(-s)\,\Gamma(-\nu-s)\left(\frac{-iz}{2}\right)^{\nu+2s} ds,$$

$$-\frac{\pi}{2} < \arg(-iz) < \frac{\pi}{2}.$$

$$2\pi^2 H_\nu^{(2)}(z) = e^{i\nu\pi/2}\int_{-e-i\infty}^{-c+i\infty}\Gamma(-s)\,\Gamma(-\nu-s)\left(\frac{iz}{2}\right)^{\nu+2s} ds,$$

$$-\frac{\pi}{2} < \arg(iz) < \frac{\pi}{2}.$$

$$2\pi i\,J_\nu(x) = \int_{-i\infty}^{i\infty}\Gamma(-s)\,\Gamma(-\nu-s)\left(-\frac{iz}{2}\right)^{\nu+2s} ds,$$

$$-\frac{\pi}{2} < \arg(-iz) < \frac{\pi}{2}.$$

Kelvin's and Associated Functions

$$\mathrm{ber}_\nu(z) + i\,\mathrm{bei}_\nu(z) = J_\nu(ze^{3\pi i/4}).$$

$$\mathrm{ber}_\nu(z) - i\,\mathrm{bei}_\nu(z) = J_\nu(ze^{-3\pi i/4}).$$

$$\mathrm{ker}_\nu(z) + i\,\mathrm{kei}_\nu(z) = K_\nu(ze^{\pi i/4}).$$

$$\mathrm{ker}_\nu(z) - i\,\mathrm{kei}_\nu(z) = K_\nu(ze^{\pi i/4}).$$

$$\mathrm{ber}(z) = \mathrm{ber}_0(z),\ \mathrm{bei}(z) = \mathrm{bei}_0(z).$$

$$\mathrm{ker}(z) = \mathrm{ker}_0(z),\ \mathrm{kei}(z) = \mathrm{kei}_0(z).$$

$$\chi_\nu^{(b)}(z) = \mathrm{ber}_\nu^2(z) + \mathrm{bei}_\nu^2(z).$$

$$V_\nu^{(b)}(z) = [\mathrm{ber}_\nu'(z)]^2 + [\mathrm{bei}_\nu'(z)]^2.$$

$$W_\nu^{(b)}(z) = \mathrm{ber}_\nu(z)\,\mathrm{bei}_\nu'(z) - \mathrm{bei}_\nu(z)\,\mathrm{ber}_\nu'(z).$$

$$\tfrac{1}{2} Z_\nu^{(n)}(z) = \mathrm{ber}_\nu(z)\,\mathrm{bei}_\nu'(z) + \mathrm{ber}_\nu(z)\,\mathrm{ber}_\nu'(z).$$

Neumann Polynomials

$$O_0(x) = \frac{1}{x};\ O_n(x) = \frac{1}{4}\sum_{r=0}^{\leqslant n/2} \frac{n(n-r+1)!}{r!\left(\dfrac{x}{2}\right)^{n-2r+1}},\ n = 1, 2, \ldots .$$

$$O_{-n}(x) = (-1)^n O_n(x),\ n = 1, 2, \ldots .$$

Anger-Weber Functions

$$\mathbf{J}_\nu(z) = \frac{1}{\pi}\int_0^\pi \cos(\nu\theta - z\sin\theta)\,d\theta$$

$$\mathbf{E}_\nu(z) = \frac{1}{\pi}\int_0^\pi \sin(\nu\theta - z\sin\theta)\,d\theta.$$

Struve's Functions

$$\mathbf{H}_\nu(z) = \sum_{r=0}^{\infty} \frac{(-1)^r\left(\dfrac{z}{2}\right)^{\nu+2r+1}}{\Gamma\left(r+\dfrac{3}{2}\right)\Gamma\left(\nu+r+\dfrac{3}{2}\right)}$$

$$= \frac{\left(\dfrac{z}{2}\right)^{\nu+1}}{\Gamma\left(\dfrac{3}{2}\right)\Gamma\left(\nu+\dfrac{3}{2}\right)}\,{}_1F_2\left(1; \frac{3}{2}, \nu+\frac{3}{2}; -\frac{z^2}{4}\right)$$

$$= \frac{2^{1-\nu}}{\Gamma\left(\frac{1}{2}\right)\Gamma\left(\nu+\frac{1}{2}\right)} s_{\mu,\nu}(z)$$

where $s_{\mu,\nu}$ is a Lommel's function.

$$\mathbf{L}_\nu(z) = \exp\left[-\frac{i\pi}{2}(\nu+1)\right] \mathbf{H}_\nu\left[z \exp\left(\frac{i\pi}{2}\right)\right]$$

$$\mathbf{H}_\nu(x) = \frac{2^{\mu-2}}{\pi i\,\Gamma\left(\frac{1}{2}-\frac{\mu}{2}\pm\frac{\nu}{2}\right)}$$

$$\times \int_{c-i\infty}^{c+i\infty} \Gamma\left(\frac{1+\mu}{2}+s\right)\Gamma\left(s+\frac{\nu}{2}\right)\Gamma\left(s-\frac{\nu}{2}\right)\Gamma\left(\frac{1-\mu}{2}-s\right)\left(\frac{x^2}{4}\right)^{-s} ds,$$

$$R(s\pm\nu)>0,\ R(s+\mu+1)>0,\ R(s)<R(1-\mu).$$

Lommel's Functions

$$s_{\mu,\nu}(z) = \frac{z^{\mu+1}}{(\mu-\nu+1)(\mu+\nu+1)}\,{}_1F_2\left(1; \frac{\mu-\nu+3}{2}, \frac{\mu+\nu+3}{2}; \frac{z^2}{4}\right)$$

$$S_{\mu,\nu}(z) = s_{\mu,\nu}(z) + 2^{\mu-1}\,\Gamma\left(\frac{\mu+\nu+1}{2}\right)\Gamma\left(\frac{\mu-\nu+1}{2}\right)$$

$$\times\left[\sin\left\{\left(\frac{\mu-\nu}{2}\right)\pi\right\} J_\nu(z) - \cos\left\{\left(\frac{\mu-\nu}{2}\right)\pi\right\} Y_\nu(z)\right]$$

Complete Elliptic Integrals

$$K(k) = \int_0^{\pi/2} (1-k^2\sin^2\phi)^{-1/2}\, d\phi$$

$$= \frac{\pi}{2}\,{}_2F_1\left(\frac{1}{2}, \frac{1}{2}; 1; k^2\right)$$

$$E(k) = \int_0^{\pi/2} (1-k^2\sin^2\phi)^{1/2}$$

$$= \frac{\pi}{2}\,{}_2F_1\left(-\frac{1}{2}, \frac{1}{2}; 1; k^2\right).$$

A5 *H*-FUNCTION AND ITS SPECIAL CASES

H-FUNCTION

$$H_{p,q}^{m\,n}\left[x \,\middle|\, \begin{matrix}(a_p, A_p)\\(b_q, B_q)\end{matrix}\right]$$

$$= \frac{1}{2\pi i} \int_{L_1} \chi(s)\, x^s\, ds.$$

where

$$\chi(s) = \frac{\prod_{j=1}^{m} \Gamma(b_j - B_j s) \prod_{j=1}^{n} \Gamma(1 - a_j + A_j s)}{\prod_{j=m+1}^{q} \Gamma(1 - b_j + B_j s) \prod_{j=n+1}^{p} \Gamma(a_j - A_j s)}$$

and L_1 is a suitable contour separating the poles of $\Gamma(b_j - B_j s)$ for $j = 1, \ldots, m$ from those of $\Gamma(1 - a_j + A_j s)$ for $j = 1, \ldots, n$. For a more detailed definition see Section 1.1.

MEIJER'S *G*-FUNCTION

$$G_{p,q}^{m,n}\left[x \,\middle|\, \begin{matrix} a_p \\ b_q \end{matrix}\right] = \frac{1}{2\pi i} \int_{L_2} \frac{\prod_{j=1}^{m} \Gamma(b_j - s) \prod_{j=1}^{n} \Gamma(1 - a_j + s)}{\prod_{j=m+1}^{q} \Gamma(1 - b_j + s) \prod_{j=n+1}^{p} \Gamma(a_j - s)}\, x^s\, ds,$$

where L_2 is a suitable contour separating the poles of $\Gamma(b_j - s)$, $(j = 1, \ldots, m)$ from those of $\Gamma(1 - a_j + s)$, $(j = 1, \ldots, n)$. For a more detailed definition and various contours, the reader is referred to the monograph by Mathai and Saxena (1973a, Section 1.1).

A 6 H-FUNCTION EXPRESSED IN TERMS OF ELEMENTARY SPECIAL FUNCTIONS

$$H_{0,1}^{1,0}[z \,|\, (\alpha, 1)] = z^{\alpha} e^{-z}.$$

$$H_{1,2}^{1,0}\left[z \,\middle|\, \begin{matrix} (\alpha, 1) \\ (\beta, 1) \end{matrix}\right] = \frac{1}{\Gamma(\alpha - \beta)} z^{\beta} (1 - z)^{\alpha - \beta - 1}, \ |z| < 1.$$

$$H_{1,2}^{1,0}\left[z \,\middle|\, \begin{matrix} (\frac{1}{2}, 1) \\ (\alpha, 1), (-\alpha, 1) \end{matrix}\right] = \pi^{-1/2} \cos(\alpha\pi)\, e^{z/2}\, I_a\left(\frac{z}{2}\right).$$

$$H_{0,2}^{1,0}[z \,|\, (\alpha, 1), (\beta, 1)] = z^{(\alpha+\beta)/2} J_{\alpha-\beta}(2z^{1/2}).$$

$$H_{0,2}^{2,0}[z \,|\, (\alpha, 1), (\beta, 1)] = 2z^{(\alpha+\beta)/2} K_{\alpha-\beta}(2z^{1/2}).$$

$$H_{1,2}^{1,1}\left[z \,\middle|\, \begin{matrix} (1/2, 1) \\ (\beta, 1), (-\beta, 1) \end{matrix}\right] = \pi^{1/2} e^{-z/2} I_{\beta}\left(\frac{z}{2}\right).$$

$$H_{1,2}^{1,1}\left[z\middle|\begin{matrix}(\alpha, 1)\\(\beta, 1), (\gamma, 1)\end{matrix}\right] = \frac{\Gamma(1-\alpha+\beta)}{\Gamma(1-\gamma+\beta)} z^{\beta} \times {}_1F_1(1-\alpha+\beta; 1+\beta-\gamma; -z).$$

$$H_{1,2}^{2,0}\left[z\middle|\begin{matrix}(1/2, 1)\\(\beta, 1), (-\beta, 1)\end{matrix}\right] = \pi^{-1/2} e^{-z/2} K_{\beta}\left(\frac{z}{2}\right).$$

$$H_{1,2}^{2,0}\left[z\middle|\begin{matrix}(\alpha, 1)\\(\beta, 1), (\gamma, 1)\end{matrix}\right] = z^{(\beta+\gamma-1)/2} e^{-z/2} W_{(1+\beta+\gamma)/2-\alpha,\ (\beta-\gamma)/2}(z).$$

$$H_{2,2}^{2,0}\left[z\middle|\begin{matrix}(\alpha+1/3, 1), (\alpha+2/3, 1)\\(\alpha, 1), (\alpha, 1)\end{matrix}\right] = z^{\alpha} \times {}_2F_1(2/3, 1/3; 1; 1-z), |z| < 1.$$

(logarithmic case)

$$H_{2,2}^{2,0}\left[z\middle|\begin{matrix}(\alpha, 1), (\alpha, 1)\\(\alpha-1/2, 1), (\alpha-1, 1)\end{matrix}\right]$$
$$= (\pi z)^{1/2} z^{\alpha-1} - 2z^{(\alpha-1/2)} \pi^{-1/2} {}_2F_1(1/2, 1/2; 3/2; z)$$
$$= \frac{1}{\Gamma(3/2)} z^{\alpha-1} (1-z)^{1/2} {}_2F_1(1/2, 1/2; 3/2; 1-z), |z| < 1.$$

(logarithmic case)

$$H_{1,2}^{2,1}\left[z\middle|\begin{matrix}(\alpha, 1)\\(\beta, 1), (\gamma, 1)\end{matrix}\right] = \Gamma(\beta-\alpha+1)\,\Gamma(\gamma-\alpha+1)\, z^{(\beta+\gamma-1)/2} \times e^{z/2} W_{\alpha-(\beta+\gamma+1)/2,\ (\beta-\gamma)/2}(z).$$

$$H_{1,2}^{2,1}\left[z\middle|\begin{matrix}(1/2, 1)\\(\alpha, 1), (-\alpha, 1)\end{matrix}\right] = \frac{\pi^{1/2}}{\cos(\alpha\pi)} e^{z/2} K_{\alpha}\left(\frac{z}{2}\right).$$

$$H_{0,4}^{1,0}[z\,|\,(\alpha, 1), (\beta, 1), (2\beta-\alpha, 1), (\beta+1/2, 1)]$$
$$= \pi^{-1/2} z^{\beta} I_{2(\alpha-\beta)}(2^{3/2} z^{1/4}) J_{2(\alpha-\beta)}(2^{3/2} z^{1/4}).$$

$$H_{0,4}^{1,0}[z\,|\,(\alpha+1/2, 1), (\alpha, 1), (\beta, 1), (2\alpha-\beta, 1)]$$
$$= \frac{z^{\alpha} \operatorname{cosec}(\alpha-\beta)\pi}{2\pi^{1/2}} [J_{2(\alpha-\beta)}(2^{3/2} z^{1/4}) I_{2(\beta-\alpha)}(2^{3/2} z^{1/4}) - I_{2(\alpha-\beta)}(2^{3/2} z^{1/4}) J_{2(\beta-\alpha)}(2^{3/2} z^{1/4})].$$

$$H_{0,4}^{1,0}[z\,|\,(\alpha, 1), (\alpha+1/2, 1), (\beta, 1), (2\alpha-\beta, 1)]$$
$$= \frac{1}{2\pi^{1/2}} \sec[(\beta-\alpha)\pi]\, z^{\alpha} [J_{2(\alpha-\beta)}(2^{3/2} z^{1/4}) J_{2(\beta-\alpha)}(2^{3/2} z^{1/4}) + I_{2(\alpha-\beta)}(2^{3/2} z^{1/4}) J_{2(\beta-\alpha)}(2^{3/2} z^{1/4})].$$

$$H_{1,3}^{1,1}\left[z\middle|\begin{matrix}(1/2, 1)\\(\alpha, 1), (0, 1) (-\alpha, 1)\end{matrix}\right] = \pi^{1/2} J_{\alpha}^{2}(z^{1/2}).$$

$$H_{1,3}^{1,1}\left[z\,\middle|\,\begin{matrix}(1/2,1)\\(0,1),(\alpha,1),(-\alpha,1)\end{matrix}\right]=\pi^{1/2}J_{\alpha}(z^{1/2})J_{-\alpha}(z^{1/2}).$$

$$H_{1,3}^{1,1}\left[z\,\middle|\,\begin{matrix}(\alpha,1)\\(\alpha,1),(\beta,1),(\alpha-\tfrac{1}{2},1)\end{matrix}\right]=z^{(\alpha+\beta)/2-1/4}H_{\alpha-\beta-1/2}(2z^{1/2}).$$

$$H_{1,3}^{2,0}\left[z\,\middle|\,\begin{matrix}(\alpha-\tfrac{1}{2},1)\\(\alpha,1),(\beta,1),(\alpha-\tfrac{1}{2},1)\end{matrix}\right]=z^{(\alpha+\beta)/2}Y_{\beta-\alpha}(2z^{1/2}).$$

$$H_{1,3}^{2,0}\left[z\,\middle|\,\begin{matrix}(\alpha+\tfrac{1}{2},1)\\(\beta,1),(\alpha,1),(2\alpha-\beta,1)\end{matrix}\right]=-\pi^{1/2}z^{\alpha}J_{\beta-\alpha}(z^{1/2})Y_{\beta-\alpha}(z^{1/2}).$$

$$H_{1,3}^{2,0}\left[z\,\middle|\,\begin{matrix}(\tfrac{1}{2},1)\\(\alpha,1),(-\alpha,1),(0,1)\end{matrix}\right]=\frac{\pi^{1/2}}{2}\operatorname{cosec}(\alpha\pi)$$
$$\times[J_{-\alpha}^{2}(z^{1/2})-J_{\alpha}^{2}(z^{1/2})].$$

$$H_{1,3}^{2,1}\left[z\,\middle|\,\begin{matrix}(\tfrac{1}{2},1)\\(\alpha,1),(0,1),(-\alpha,1)\end{matrix}\right]=2\pi^{1/2}I_{\alpha}(z^{1/2})K_{\alpha}(z^{1/2}).$$

$$H_{1,3}^{2,1}\left[z\,\middle|\,\begin{matrix}(\tfrac{1}{2},1)\\(\alpha,0),(-\alpha,0),(0,1)\end{matrix}\right]=\pi^{3/2}\operatorname{cosec}(2\alpha\pi)$$
$$\times[I_{-\alpha}^{2}(z^{1/2})-I_{\alpha}^{2}(z^{1/2})].$$

$$H_{1,3}^{2,1}\left[z\,\middle|\,\begin{matrix}(\alpha+\tfrac{1}{2},1)\\(\alpha+\tfrac{1}{2},1),(\beta,1),(\alpha,1)\end{matrix}\right]$$
$$=\frac{\pi z^{(\alpha+\beta)/2}}{\cos(\alpha-\beta)\pi}[I_{\beta-\alpha}(2z^{1/2})-\mathbf{L}_{\alpha-\beta}(2z^{1/2})].$$

$$H_{1,3}^{2,1}\left[z\,\middle|\,\begin{matrix}(\alpha+\tfrac{1}{2},1)\\(\alpha,1),(\alpha+\tfrac{1}{2},1),(\beta,1)\end{matrix}\right]$$
$$=\pi z^{(\alpha+\beta)/2}[I_{\alpha-\beta}(2z^{1/2})-\mathbf{L}_{\alpha-\beta}(2z^{1/2})].$$

$$H_{0,4}^{2,0}[z\,|\,(\alpha,1),(\alpha+\tfrac{1}{2},1),(\beta,1),(\beta+\tfrac{1}{2},1)]=z^{(\alpha+\beta)/2}J_{2(\alpha-\beta)}(4z^{1/4}).$$

$$H_{0,4}^{0,2}[z\,|\,(\alpha,1),(-\alpha,1),(0,1),(\tfrac{1}{2},1)]=-\pi^{1/2}\operatorname{cosec}(2\alpha\pi)$$
$$\times[J_{2\alpha}(\omega e^{i\pi/4})J_{2\alpha}(\omega e^{-i\pi/4})-J_{2\alpha}(\omega e^{i\pi/4})J_{-2\alpha}(\omega e^{-i\pi/4})],$$
$$\omega=2^{3/2}z^{1/4}.$$

$$H_{0,4}^{2,0}[z\,|\,(0,1),(\tfrac{1}{2},1),(\alpha,1),(-\alpha,1)]$$
$$=-\pi^{1/2}\operatorname{cosec}(2\alpha\pi)[\exp(2\alpha\pi i)J_{2\alpha}(\omega e^{-\pi i/4})J_{-2\alpha}(\omega e^{\pi i/4})]$$
$$-\exp(-2\alpha\pi i)J_{2\alpha}(\omega e^{\pi i/4})J_{-2\alpha}(\omega e^{-\pi i/4}),\ \omega=2^{3/2}z^{1/4}.$$

$$H_{2,2}^{2,0}\left[z\,\middle|\,\begin{matrix}(\alpha-1,1),(\beta-1,1)\\(\gamma-1,1),(\delta-1,1)\end{matrix}\right]=\frac{z^{\delta}(1-z)^{\alpha+\beta-\gamma-\delta-1}}{\Gamma(\alpha+\beta-\gamma-\delta)}$$
$$\times\,{}_2F_1(\alpha-\gamma,\beta-\gamma;\alpha+\beta-\gamma-\delta;1-z),\ |z|<1.$$

$$H_{1,3}^{3,0}\left[\frac{z}{9}\,\middle|\,\begin{matrix}(3a,3)\\(a,1),(a-\tfrac{1}{2},1),(a-1,1)\end{matrix}\right]$$
$$=2\pi z\,3^{-3a+1/2}\,G_{2,2}^{2,0}\left(z\,\middle|\,\begin{matrix}a+\tfrac{2}{3},\ a+\tfrac{1}{3}\\a-\tfrac{3}{2},\ a-2\end{matrix}\right)$$
$$=8\pi^{1/2}\,3^{-3a-1/2}\,z^{a-1}(1-z)^{3/2}\,{}_2F_1(\tfrac{5}{6},\tfrac{7}{6},\tfrac{5}{2};1-z),\ 0<|z|<1.$$

$$H_{0,3}^{3,0}\left(z\,\middle|\,\begin{matrix}(\alpha+1,1),(\alpha+\tfrac{3}{2},1),(\alpha+1,1)\\(\alpha,1),(\alpha-\tfrac{1}{2},1),(\alpha-1,1)\end{matrix}\right)$$
$$=4z^{a-1}\left[-2z^{1/2}+\tfrac{2}{3}z^{3/2}+\tfrac{1}{3}z(1-\log z)\right],\ |z|<1.$$

$$H_{1,3}^{3,0}\left[z\,\middle|\,\begin{matrix}(\alpha+\tfrac{1}{2},1)\\(\alpha+\beta,1),(\alpha-\beta,1),(\alpha,1)\end{matrix}\right]=2\pi^{-1/2}\,z^{\alpha}K_{\beta}^{2}(z^{1/2}).$$

$$H_{1,3}^{3,1}\left[z\,\middle|\,\begin{matrix}(\alpha+\tfrac{1}{2},1)\\(\alpha+\tfrac{1}{2},1),(-\alpha,1),(\alpha,1)\end{matrix}\right]$$
$$=\frac{\pi^2}{\cos(2\alpha\pi)}\left[H_{2\alpha}(2z^{1/2})-Y_{2\alpha}(2z^{1/2})\right].$$

$$H_{1,3}^{3,1}\left[z\,\middle|\,\begin{matrix}(\alpha,1)\\(\alpha,1),(\beta,1),(-\beta,1)\end{matrix}\right]$$
$$=2^{2-2\alpha}\,\Gamma(1-\alpha-\beta)\,\Gamma(1-\alpha+\beta)\,S_{2\alpha-1,\,2\beta}(2z^{1/2}).$$

$$H_{1,3}^{3,1}\left[z\,\middle|\,\begin{matrix}(\alpha+\tfrac{1}{2},1)\\(\beta,1),(2\alpha-\beta,1),(\alpha,1)\end{matrix}\right]$$
$$=\frac{1}{2}\,\pi^{5/2}\sec(\beta-\alpha)\pi\,z^{\alpha}\,H_{\beta-\alpha}^{(1)}(z^{1/2})\,H_{\beta-\alpha}^{(1)}(z^{1/2}).$$

$$H_{0,4}^{3,0}\left[z\,|\,(3\alpha-\tfrac{1}{2},1),(\alpha,1),(-\alpha-\tfrac{1}{2},1),(\alpha-\tfrac{1}{2},1)\right]$$
$$=2\pi^{1/2}\sec(2\alpha\pi)\,z^{\alpha-1/2}\,K_{4\alpha}(2^{3/2}z^{1/4})$$
$$\times\left[J_{4\alpha}(2^{3/2}z^{1/4})+J_{-4\alpha}(2^{3/2}z^{1/4})\right].$$

$$H_{0,4}^{3,0}\left[z\,|\,(0,1),(\alpha-\tfrac{1}{2},1),(-\alpha-\tfrac{1}{2},1),(-\tfrac{1}{2},1)\right]$$
$$=4\pi^{1/2}z^{-1/2}K_{2\alpha}(2^{3/2}z^{1/4})\left[J_{2\alpha}(2^{3/2}z^{1/4})\cos(\alpha\pi)\right.$$
$$\left.-\,Y_{2\alpha}(2^{3/2}z^{1/4})\sin(\alpha\pi)\right].$$

$$H_{0,4}^{3,0}\left[z\,|\,(-\tfrac{1}{2},1),(\alpha-\tfrac{1}{2},1),(-\alpha-\tfrac{1}{2},1),(0,1)\right]$$
$$=-4\pi^{1/2}z^{-1/2}K_{2\alpha}(2^{3/2}z^{1/4})\left[J_{2\alpha}(2^{3/2}z^{1/4})\sin(\alpha\pi)\right.$$
$$\left.+\,Y_{2\alpha}(2^{3/2}z^{1/4})\cos(\alpha\pi)\right].$$

$$H^{3,0}_{0,4}\,[z \mid (\alpha, 1), (\beta + \tfrac{1}{2}, 1), (\beta, 1), (2\beta - \alpha, 1)]$$
$$= (2\pi)^{1/2}\, z^{\beta}\, K_{2(\alpha-\beta)}\,(2^{3/2}\, z^{1/4})\, J_{2(\alpha-\beta)}\,(2^{3/2}\, z^{1/4}).$$

$$H^{4,0}_{0,4}\,[z \mid (\alpha, 1), (\alpha + \tfrac{1}{2}, 1), (\beta, 1), (\beta + \tfrac{1}{2}, 1)]$$
$$= 4\pi z^{(\alpha+\beta)/2}\, K_{2(\alpha-\beta)}\,(4z^{1/4}).$$

$$H^{4,0}_{0,4}\,[z \mid (\alpha, 1), (\alpha + \tfrac{1}{2}, 1), (\beta, 1), (2\alpha - \beta, 1)]$$
$$= 2^{3/2}\, \pi^{1/2}\, z^{\alpha}\, K_{2(\beta-\alpha)}\,(2^{3/2}\, z^{1/4}\, e^{\pi i/4})\, K_{2(\beta-\alpha)}\,(2^{3/2}\, z^{1/4}\, e^{-\pi i/4}).$$

$$H^{n,0}_{0,4}\,[z \mid (\alpha, 1), (\beta, 1), (\gamma, 1), (\delta, 1)] = z^{-1/2}\, S_n\,(\alpha, \beta, \gamma, \delta; z^{1/2}),$$

where $n = 1, 2, 3, 4$.

$$H^{1,2}_{2,2}\left[z \,\middle|\, \begin{matrix}(\alpha, 1), (\beta, 1)\\ (\gamma, 1), (\delta, 1)\end{matrix}\right]$$
$$= \frac{\Gamma(\gamma-\alpha+1)\,\Gamma(\gamma-\beta+1)}{\Gamma(\gamma-\delta+1)}\, z^{\gamma}\, {}_2F_1\,(\gamma-\alpha+1, \gamma-\beta+1; \gamma-\delta+1; -z).$$

$$H^{1,2}_{2,4}\left[z \,\middle|\, \begin{matrix}(\alpha+\tfrac{1}{2}, 1), (\alpha, 1)\\ (\alpha+\beta, 1), (\alpha-\gamma, 1), (\alpha+\gamma, 1), (\alpha-\beta, 1)\end{matrix}\right]$$
$$= \pi^{1/2}\, z^{\alpha}\, J_{\beta+\gamma}\,(z^{1/2})\, J_{\beta-\gamma}\,(z^{1/2}).$$

$$H^{2,2}_{2,4}\left[z \,\middle|\, \begin{matrix}(\alpha, 1), (\alpha + \tfrac{1}{2}, 1)\\ (\beta, 1), (\gamma, 1), (2\alpha - \gamma, 1), (2\alpha - \beta, 1)\end{matrix}\right]$$
$$= 2\pi^{1/2}\, z^{\alpha}\, I_{\beta+\gamma-2\alpha}\,(z^{1/2})\, K_{\beta-\gamma}\,(z^{1/2}).$$

$$H^{3,0}_{2,4}\left[z \,\middle|\, \begin{matrix}(0, 1), (\tfrac{1}{2}, 1)\\ (\alpha, 1), (\beta, 1), (-\beta, 1), (-\alpha, 1)\end{matrix}\right]$$
$$= \frac{i}{4}\pi^{1/2}\left[H^{(1)}_{\alpha-\beta}\,(z^{1/2})\, H^{(1)}_{\alpha+\beta}\,(z^{1/2}) - H^{(2)}_{\alpha-\beta}\,(z^{1/2})\, H^{(2)}_{\alpha+\beta}\,(z^{1/2})\right].$$

$$H^{3,1}_{2,4}\left[z \,\middle|\, \begin{matrix}(\tfrac{1}{2} + \alpha, 1), (\tfrac{1}{2} - \alpha, 1)\\ (0, 1), (\tfrac{1}{2}, 1), (\beta, 1), (-\beta, 1)\end{matrix}\right]$$
$$= \frac{\pi^{1/2}\,\Gamma(\tfrac{1}{2} - \alpha + \beta)}{z^{1/2}\,\Gamma(1 + 2\alpha)}\, W_{\alpha,\beta}\,(2z^{1/2})\, M_{-\alpha,\beta}\,(2z^{1/2}).$$

$$H^{4,0}_{2,4}\left[z \,\middle|\, \begin{matrix}(\tfrac{1}{2} + \alpha, 1), (\tfrac{1}{2} - \alpha, 1)\\ (0, 1), (\tfrac{1}{2}, 1), (\beta, 1), (-\beta, 1)\end{matrix}\right] = \pi^{1/2}\, z^{-1/2}\, W_{\alpha,\beta}\,(2z^{1/2})$$
$$\times W_{-\alpha,\beta}\,(2z^{1/2}).$$

$$H_{2,4}^{4,0}\left[z\,\middle|\,\begin{matrix}(\alpha, 1), (\alpha+\frac{1}{2}, 1)\\(\beta+\gamma, 1), (\beta-\gamma, 1), (\beta+\gamma+\frac{1}{2}, 1), (\beta-\gamma+\frac{1}{2}, 1)\end{matrix}\right]$$
$$=\pi^{1/2}\, 2^{2\alpha-2\beta-1/2}\, z^{\beta-1/4}\, e^{-\sqrt{z}}\, W_{1/2+2\beta-2\alpha, 2\gamma}(2z^{1/2}).$$

$$H_{2,4}^{4,0}\left[z\,\middle|\,\begin{matrix}(\alpha, 1), (\alpha+\frac{1}{2}, 1)\\(\alpha+\beta, 1), (\alpha+\gamma, 1), (\alpha-\gamma, 1), (\alpha-\beta, 1)\end{matrix}\right]$$
$$=2\pi^{-1/2}\, z^{\alpha}\, K_{\beta+\gamma}(z^{1/2})\, K_{\beta-\gamma}(z^{12/}).$$

$$H_{2,4}^{4,1}\left[z\,\middle|\,\begin{matrix}(0, 1), (\frac{1}{2}, 1)\\(\alpha, 1), (\beta, 1), (-\alpha, 1), (-\beta, 1)\end{matrix}\right]$$
$$=\frac{i\pi^{5/2}}{4\sin(\alpha\pi)\sin(\beta\pi)}\Big[\exp(-\beta\pi i)\, H_{\alpha-\beta}^{(1)}(z^{1/2})\, H_{\alpha+\beta}^{(2)}(z^{1/2})$$
$$-\exp(\beta\pi i)\, H_{\alpha+\beta}^{(1)}(z^{1/2})\, H_{\alpha-\beta}^{(2)}(z^{1/2})\Big].$$

$$H_{2,4}^{4,1}\left[z\,\middle|\,\begin{matrix}(\frac{1}{2}, 1), (0, 1)\\(\alpha, 1), (\beta, 1), (-\beta, 1), (-\alpha, 1)\end{matrix}\right]$$
$$=\frac{\pi^{5/2}}{4\cos(\alpha\pi)\cos(\beta\pi)}\Big[\exp(-\beta\pi i)\, H_{\alpha-\beta}^{(1)}(z^{1/2})\, H_{\alpha+\beta}^{(2)}(z^{1/2})$$
$$+\exp(\beta\pi i)\, H_{\alpha+\beta}^{(1)}(z^{1/2})\, H_{\alpha-\beta}^{(2)}(z^{1/2})\Big].$$

$$H_{2,4}^{4,1}\left[z\,\middle|\,\begin{matrix}(\frac{1}{2}+\alpha, 1), (\frac{1}{2}-\alpha, 1)\\(0, 1), (\frac{1}{2}, 1), (\beta, 1), (-\beta, 1)\end{matrix}\right]$$
$$=z^{-1/2}\,\pi^{1/2}\,\Gamma(\tfrac{1}{2}-\alpha+\beta)\,\Gamma(\tfrac{1}{2}-\alpha-\beta)$$
$$\times W_{\alpha,\beta}(2iz^{1/2})\, W_{\alpha,\beta}(-2iz^{1/2}).$$

$$H_{2,4}^{4,2}\left[z\,\middle|\,\begin{matrix}(\alpha, 1), (\alpha+\frac{1}{2}, 1)\\(\beta+\gamma, 1), (\beta-\gamma, 1), (\beta+\gamma+\frac{1}{2}, 1), (\beta-\gamma+\frac{1}{2}, 1)\end{matrix}\right]$$
$$=2^{2\alpha-2\beta+1/2}\,\pi^{3/2}\,\Gamma(1-2\alpha+2\beta+2\gamma)\,\Gamma(1-2\alpha+2\beta-2\gamma)$$
$$\times z^{\beta-1/4}\, e^{\sqrt{z}}\, W_{2\alpha-2\beta-1/2, 2\gamma}(2z^{1/2}).$$

$$H_{4,4}^{4,0}\left[z\,\middle|\,\begin{matrix}(\alpha, 1), (\alpha+\frac{1}{2}, 1), (\alpha, 1), (\alpha+\frac{1}{2}, 1)\\(\alpha-1, 1), (\alpha-\frac{3}{2}, 1), (\alpha-2, 1), \alpha-\frac{5}{2}, 1)\end{matrix}\right]$$
$$=z^{\alpha-2}\,[a_1 z^{-2}+a_2(b_2-\log z)\, z^{-1}$$
$$+c_1 z^{-5/2}+c_2(d_2-\log z)\, z^{-3/2}+c_3(d_3-\log z)\, z^{-1/2}],$$

where

$$a_1=-\frac{2^5}{9},\ a_2=\frac{2^5}{3},\ b_2=\frac{5}{3},\ c_1=\frac{2}{9},\ c_2=8,\ c_3=\frac{4}{3},\ d_2=-3$$

and

$$d_3=\frac{43}{6}$$

$$H^{1,p}_{p,q}\left[z\,\middle|\,\begin{matrix}(\alpha_p, 1)\\(\beta_q, 1)\end{matrix}\right]=\frac{\prod_{j=1}^{p}\Gamma(1+\beta_1-\alpha_j)\,z^{\beta_1}}{\prod_{j=2}^{q}\Gamma(1+\beta_1-\beta_j)}$$

$$\times\,{}_pF_{q-1}\left[\begin{matrix}1+\beta_1-\alpha_1,\ldots,1+\beta_1-\alpha_p\\1+\beta_1-\beta_2,\ldots,1+\beta_1-\beta_q\end{matrix};-z\right],\ p\leqslant q.$$

$$H^{1,n}_{p,q}\left[z\,\middle|\,\begin{matrix}(a_p, 1)\\(b_q, 1)\end{matrix}\right]=\frac{\prod_{j=1}^{n}\Gamma(1+b_1-a_j)\,z^{b_1}}{\prod_{j=2}^{q}\Gamma(1+b_1-b_j)\prod_{j=n+1}^{p}\Gamma(a_j-b_1)}$$

$$\times\,{}_pF_{q-1}\left(\begin{matrix}1+b_1-a_1,\ldots,1+b_1-a_p\\1+b_1-b_2,\ldots,1+b_1-b_q\end{matrix};(-1)^{p-n-1}z\right),\ p\leqslant q.$$

$$H^{1,p}_{p,q+1}\left[z\,\middle|\,\begin{matrix}(1-a_p, A_p)\\(0,1),(1-b_q, B_q)\end{matrix}\right]$$

$$={}_p\psi_q\left[\begin{matrix}(a_p, A_p)\\(b_q, B_q)\end{matrix};-z\right]=\sum_{r=0}^{\infty}\frac{\prod_{j=1}^{p}\Gamma(a_j+A_jr)(-z)^r}{\prod_{j=1}^{q}\Gamma(b_j+B_jr)\,r!},$$

where ${}_p\psi_q$ is the Maitland's generalized hypergeometric function.

$$H^{1,0}_{0,2}\,[z\,|\,(0,1),(-\nu,\mu)]=J^{\mu}_{\nu}(z)=\sum_{r=0}^{\infty}\frac{(-z)^r}{r!\;\Gamma(1+\nu+\mu r)}.$$

A7 ELEMENTARY SPECIAL FUNCTIONS EXPRESSED IN TERMS OF H-FUNCTIONS

$$z^{\alpha}\,e^{-z}=H^{1,0}_{0,1}\,(z\,|\,(\alpha,1))$$

$$\sin z=\left(\frac{\pi z}{2}\right)^{1/2}H^{1,0}_{0,2}\left[\frac{z^2}{4}\,\middle|\,\left(\frac{1}{4},1\right),\left(-\frac{1}{4},1\right)\right]$$

$$=\pi^{1/2}\,H^{1,0}_{0,2}\left(\frac{z^2}{4}\,\middle|\,\left(\frac{1}{2},1\right),(0,1)\right)$$

$$\cos z=\left(\frac{\pi z}{2}\right)^{1/2}H^{1,0}_{0,2}\left[\frac{z^2}{4}\,\middle|\,\left(-\frac{1}{4},1\right),\left(\frac{1}{4},1\right)\right]$$

$$=\pi^{1/2}\,H^{1,0}_{0,2}\left[\frac{z^2}{4}\,\middle|\,(0,1),\left(\frac{1}{2},1\right)\right]$$

$$\sinh z=\frac{\pi^{1/2}}{i}\,H^{1,0}_{0,2}\left[-\frac{z^2}{4}\,\middle|\,\left(\frac{1}{2},1\right),(0,1)\right]$$

$$\cosh z = \pi^{1/2} \, H_{0,2}^{1,0}\left[-\frac{z^2}{4}\,\bigg|\,(0,1),\,\left(\frac{1}{2},1\right)\right]$$

$$\log(1 \pm z) = \pm\, H_{2,2}^{1,0}\left[\pm z\,\bigg|\,\begin{matrix}(1,1),\,(1,1)\\(1,1),\,(0,1)\end{matrix}\right]$$

$$\log\left(\frac{1+z}{1-z}\right) = z\, H_{2,2}^{1,2}\left[-z^2\,\bigg|\,\begin{matrix}(\frac{1}{2},1),\,(0,1)\\(0,1),\,(\frac{1}{2},1)\end{matrix}\right]$$

$$\arcsin z = \tfrac{1}{2}\, H_{2,2}^{1,2}\left[-z^2\,\bigg|\,\begin{matrix}(\frac{1}{2},1),\,(\frac{1}{2},1)\\(0,1),\,(-\frac{1}{2},1)\end{matrix}\right]$$

$$\arctan z = \tfrac{1}{2}\, H_{2,2}^{1,2}\left[z^2\,\bigg|\,\begin{matrix}(1,1),\,(\frac{1}{2},1)\\(\frac{1}{2},1),\,(0,1)\end{matrix}\right]$$

$$\log(z + \sqrt{1+z^2}) = \frac{1}{2\pi^{1/2}}\, H_{2,2}^{1,2}\left[z^2\,\bigg|\,\begin{matrix}(1,1),\,(1,1)\\(\frac{1}{2},1),\,(0,1)\end{matrix}\right]$$

$$(1+z^2)^{-1/2}[1+(1+z^2)^{1/2}]^{-2\alpha} = \pi^{-1/2}\, H_{2,2}^{1,2}\left[z^2\,\bigg|\,\begin{matrix}(-\alpha,1),\,(\frac{1}{2}-\alpha,1)\\(0,1),\,(-2\alpha,1)\end{matrix}\right]$$

$$(1+z)^{-2\alpha} + (1-z)^{-2\alpha} = \frac{1}{2^{2\alpha}\,\Gamma(2\alpha)}\, H_{2,2}^{1,2}\left[-z^2\,\bigg|\,\begin{matrix}(1-\alpha,1),\,(\frac{1}{2}-\alpha,1)\\(0,1),\,(\frac{1}{2},1)\end{matrix}\right]$$

$$(1+z)^{1-2\alpha} - (1-z)^{1-2\alpha} = -\frac{z}{2^{2\alpha-1}\,\Gamma(2\alpha-1)}$$
$$H_{2,2}^{1,2}\left[-z^2\,\bigg|\,\begin{matrix}(1-\alpha,1),\,(\frac{1}{2}-\alpha,1)\\(0,1),\,(-\frac{1}{2},1)\end{matrix}\right]$$

$$z^\alpha(1-z)^\beta = \Gamma(\beta+1)\, H_{1,1}^{1,0}\left[z\,\bigg|\,\begin{matrix}(\alpha+\beta+1,1)\\(\alpha,1)\end{matrix}\right].$$

$$\frac{z^\beta}{1+az^\alpha} = a^{-\beta/\alpha}\, H_{1,1}^{1,1}\left[az^\alpha\,\bigg|\,\begin{matrix}(\beta/\alpha,1)\\(\beta/\alpha,1)\end{matrix}\right].$$

$$(1-z)^{\alpha-1} = \Gamma(\alpha)\, H_{1,1}^{1,1}\left[z\,\bigg|\,\begin{matrix}(\alpha,1)\\(0,1)\end{matrix}\right].$$

$$(1-z)^{-\alpha} = \frac{1}{\Gamma(\alpha)}\, H_{1,1}^{1,1}\left[-z\,\bigg|\,\begin{matrix}(1-\alpha,1)\\(0,1)\end{matrix}\right].$$

$$z^\alpha J_\nu(z) = 2^\alpha\, H_{0,2}^{1,0}\left[\frac{z^2}{4}\,\bigg|\,\left(\frac{\alpha+\nu}{2},1\right),\,\left(\frac{\alpha-\nu}{2},1\right)\right]$$
$$= 2^{2\alpha}\, H_{0,4}^{2,0}\left[\frac{z^4}{256}\,\bigg|\,\left\{\Delta\left(2,\frac{\alpha+\nu}{2}\right),1\right\},\,\left\{\Delta\left(2,\frac{\alpha-\nu}{2}\right)\,1\right\}\right].$$

$$z^{\alpha} K_{\nu}(z) = 2^{\alpha-1} H^{2,0}_{0,2}\left[\frac{z^2}{4}\middle|\left(\frac{\alpha+\nu}{2}, 1\right), \left(\frac{\alpha-\nu}{2}, 1\right)\right]$$

$$= 2^{2\alpha-2}\pi^{-1}$$

$$\times H^{4,0}_{0,4}\left[\frac{z^4}{256}\middle|\left\{\Delta\left(2, \frac{\alpha+\nu}{2}\right), 1\right\}, \left\{\Delta\left(2, \frac{\alpha-\nu}{2}\right), 1\right\}\right].$$

$$z^{\alpha} Y_{\nu}(z) = 2^{\alpha} H^{2,0}_{1,3}\left[\frac{z^2}{4}\middle|\begin{matrix}\left(\frac{\alpha-\nu-1}{2}, 1\right)\\ \left(\frac{\alpha+\nu}{2}, 1\right), \left(\frac{\alpha-\nu}{2}, 1\right), \left(\frac{\alpha-\nu-1}{2}, 1\right)\end{matrix}\right].$$

$$(2z)^{\alpha} e^{-z} I_{\nu}(z) = \pi^{-1/2} H^{1,1}_{1,2}\left[2z\middle|\begin{matrix}(\frac{1}{2}+\alpha, 1)\\ (\alpha+\nu, 1), (\alpha-\nu, 1)\end{matrix}\right].$$

NOTE: The symbol $\left\{\Delta\left(2, \frac{\alpha+\nu}{2}\right), 1\right\}$ denotes the parameters

$$\left(\frac{\alpha+\nu}{4}, 1\right) \text{ and } \left(\frac{\alpha+\nu+2}{4}, 1\right).$$

$$(2z)^{\alpha} e^{-z} K_{\nu}(z) = \pi^{1/2} H^{2,0}_{1,2}\left[2z\middle|\begin{matrix}(\frac{1}{2}+\alpha, 1)\\ (\alpha+\nu, 1), (\alpha-\nu, 1)\end{matrix}\right]$$

$$(2z)^{\alpha} e^{z} K\nu(z) = \pi^{-1/2}\cos(\nu\pi)\, H^{2,1}_{1,2}\left[2z\middle|\begin{matrix}(\frac{1}{2}+\alpha, 1)\\ (\alpha+\nu, 1), (\alpha-\nu, 1)\end{matrix}\right]$$

$$\left(\frac{z}{2}\right)^{\alpha} s_{\mu,\nu}(z) = 2^{\mu-1}\Gamma\left(\frac{\mu+\nu+1}{2}\right)\Gamma\left(\frac{\mu-\nu+1}{2}\right)$$

$$\times H^{1,1}_{1,3}\left[\frac{z^2}{4}\middle|\begin{matrix}\left(\frac{\alpha+\mu+\nu}{2}, 1\right)\\ \left(\frac{\alpha+\mu+1}{2}, 1\right), \left(\frac{\alpha+\nu}{2}, 1\right), \left(\frac{\alpha-\nu}{2}, 1\right)\end{matrix}\right]$$

$$\left(\frac{z}{2}\right)^{\alpha} \mathbf{H}_{\nu}(z) = H^{1,1}_{1,3}\left[\frac{z^2}{4}\middle|\begin{matrix}\left(\frac{\alpha+\nu+1}{2}, 1\right)\\ \left(\frac{\alpha+\nu+1}{2}, 1\right), \left(\frac{\alpha+\nu}{2}, 1\right), \left(\frac{\alpha-\nu}{2}, 1\right)\end{matrix}\right].$$

$$\left(\frac{z}{2}\right)^{\alpha} [\mathbf{H}_{\nu}(z) - Y_{\nu}(z)]$$

$$= \frac{\cos(\nu\pi)}{\pi^2} H^{3,1}_{1,3}\left[\frac{z^2}{4}\middle|\begin{matrix}\left(\frac{\alpha+\nu+1}{2}, 1\right)\\ \left(\frac{\alpha+\nu+1}{2}, 1\right), \left(\frac{\alpha+\nu}{2}, 1\right), \left(\frac{\alpha-\nu}{2}, 1\right)\end{matrix}\right]$$

$$\left(\frac{z}{2}\right)^{\alpha}[I_{\nu}(z)-\mathbf{L}_{\nu}(z)] = \frac{1}{\pi}H_{1,3}^{2,1}\left[\frac{z^2}{4}\middle|\begin{matrix}\left(\frac{\alpha+\nu+1}{2},1\right)\\ \left(\frac{\alpha+\nu+1}{2},1\right),\left(\frac{\alpha+\nu}{2},1\right),\left(\frac{\alpha-\nu}{2},1\right)\end{matrix}\right].$$

$$\left(\frac{z}{2}\right)^{\alpha}[I_{-\nu}(z)-\mathbf{L}_{\nu}(z)] = \pi^{-1}\cos(\nu\pi)\,H_{1,3}^{2,1}\left[\frac{z^2}{4}\middle|\begin{matrix}\left(\frac{\alpha+\nu+1}{2},1\right)\\ \left(\frac{\alpha+\nu+1}{2},1\right),\left(\frac{\alpha-\nu}{2},1\right),\left(\frac{\alpha+\nu}{2},1\right)\end{matrix}\right].$$

$$\left(\frac{z}{2}\right)^{\alpha}S_{\mu,\nu}(z) = \frac{2^{\mu-1}}{\Gamma\left(\frac{1-\mu+\nu}{2}\right)\Gamma\left(\frac{1-\mu-\nu}{2}\right)} \times H_{1,3}^{3,1}\left[\frac{z^2}{4}\middle|\begin{matrix}\left(\frac{\alpha+\mu+1}{2},1\right)\\ \left(\frac{\alpha+\mu+1}{2},1\right),\left(\frac{\alpha+\nu}{2},1\right),\left(\frac{\alpha-\nu}{2},1\right)\end{matrix}\right].$$

$$Ci(z)+i\,Si(z) = -\frac{z^{-1}e^{iz}}{\pi^{1/2}}\left\{2z^{-1}\,H_{1,3}^{3,1}\left[\frac{z^2}{4}\middle|\begin{matrix}(1,1)\\ (\tfrac{3}{2},1),\ (1,1),\ (1,1)\end{matrix}\right] + i\,H_{1,3}^{3,1}\left[\frac{z^2}{4}\middle|\begin{matrix}(1,1)\\ (\tfrac{1}{2},1),\ (1,1),\ (1,1)\end{matrix}\right]\right\}.$$

$$C(z)+i\,S(z) = \frac{1+i}{2} - \frac{(2\pi z)^{-1/2}e^{iz}}{2^{1/2}}\left\{\frac{2}{z}H_{1,3}^{3,1}\left[\frac{z^2}{4}\middle|\begin{matrix}(1,1)\\ (\tfrac{3}{4},1),\ (\tfrac{5}{4},1),\ (1,1)\end{matrix}\right] + i\,H_{1,3}^{3,1}\left[\frac{z^2}{4}\middle|\begin{matrix}(1,1)\\ (\tfrac{1}{4},1),\ (\tfrac{3}{4},1),\ (1,1)\end{matrix}\right]\right\}.$$

$$z^{\sigma}e^{z/2}M_{k,m}(z) = z^{1/2}\,\Gamma(2m+1)\,\Gamma(\tfrac{1}{2}+k-m)\,H_{1,2}^{1,0}\left[z\middle|\begin{matrix}(\tfrac{1}{2}+\sigma+k,1)\\ (\sigma+m,1),\ \sigma-m,1)\end{matrix}\right]$$

$$= \frac{(2\pi z)^{1/2}\,\Gamma(2m+1)\,\Gamma(k-m+\tfrac{1}{2})}{2^{k-\sigma}} \times H_{2,4}^{2,0}\left[\frac{z^2}{4}\middle|\begin{matrix}\{\Delta(2,\tfrac{1}{2}+\sigma+k),1\}\\ \{\Delta(2,\sigma+m),1\},\ \{\Delta(2,\sigma-m),1\}\end{matrix}\right].$$

$$z^{\sigma}e^{-z/2}M_{k,m}(z) = \frac{z^{1/2}\,\Gamma(2m+1)}{\Gamma(\tfrac{1}{2}+k+m)}\,H_{1,2}^{1,1}\left[z\middle|\begin{matrix}(\tfrac{1}{2}+\sigma-k,1)\\ (\sigma+m,1),(\sigma-m,1)\end{matrix}\right]$$

$$= \frac{(z/2\pi)^{1/2}\,\Gamma(2m+1)\,2^{k+\sigma}}{\Gamma(\tfrac{1}{2}+k+m)} \times H_{2,4}^{2,2}\left[\frac{z^2}{4}\middle|\begin{matrix}\{\Delta(2,\tfrac{1}{2}+\sigma-k),1\}\\ \{\Delta(2,\sigma+m),1\},\ \{\Delta(2,\sigma-m),1\}\end{matrix}\right].$$

$$z^{\rho}\, e^{z/2}\, W_{\lambda,\mu}(z) = \frac{1}{\Gamma(\frac{1}{2}-\lambda+\mu)\,\Gamma(\frac{1}{2}-\lambda-\mu)} \times H^{2,1}_{1,2}\left[z \middle| \begin{matrix} (1+\rho+\lambda,\,1) \\ (\frac{1}{2}+\mu+\rho,\,1),\,(\frac{1}{2}-\mu+\rho,\,1) \end{matrix}\right]$$

$$= \frac{z^{1/2}\, 2^{\rho-\lambda}}{(2\pi)^{3/2}\,\Gamma(\frac{1}{2}-\lambda+\mu)\,\Gamma(\frac{1}{2}-\lambda-\mu)} \times H^{4,2}_{2,4}\left[\frac{z^2}{4} \middle| \begin{matrix} \{\Delta(2,\,\frac{1}{2}+\lambda+\rho),\,1\} \\ \{\Delta(2,\,\rho+\mu),\,1\},\,\{\Delta(2,\,\rho-\mu),\,1\} \end{matrix}\right].$$

$$z^{\rho}\, e^{-z/2}\, W_{\lambda,\mu}(z) = z^{1/2}\; H^{2,0}_{1,2}\left[z \middle| \begin{matrix} (\frac{1}{2}+\rho-\lambda,\,1) \\ (\rho+\mu,\,1),\,(\rho-\mu,\,1) \end{matrix}\right]$$

$$= \left(\frac{z}{2\pi}\right)^{1/2} 2^{\rho+\lambda}\; H^{4,0}_{2,4}\left[\frac{z^2}{4} \middle| \begin{matrix} \{\Delta(2,\,\frac{1}{2}+\rho-\lambda),\,1\} \\ \{\Delta(2,\,\rho+\mu),\,1\},\,\{\Delta(2,\,\rho-\mu),\,1\} \end{matrix}\right].$$

$$W_{\lambda,\mu}(z)\, M_{-\lambda,\mu}(z) = \frac{\Gamma(1+2\mu)}{\Gamma(\frac{1}{2})\,\Gamma(\frac{1}{2}-\lambda+\mu)} \times H^{3,1}_{2,4}\left[\frac{z^2}{4} \middle| \begin{matrix} (1+\lambda,\,1),\,(1-\lambda,\,1) \\ (\frac{1}{2},\,1),\,(1,\,1),\,(\frac{1}{2}+\mu,\,1),\,(\frac{1}{2}-\mu,\,1) \end{matrix}\right].$$

$$z^{\rho}\, W_{\lambda,\mu}(2iz)\, W_{\lambda,\mu}(-2iz) = \frac{z}{\Gamma(\frac{1}{2})\,\Gamma(\frac{1}{2}-\lambda+\mu)\,\Gamma(\frac{1}{2}-\lambda-\mu)} \times H^{4,1}_{2,4}\left[z^2 \middle| \begin{matrix} \left(\frac{1}{2}+\frac{\rho}{2}+\lambda,\,1\right),\,\left(\frac{1}{2}+\frac{\rho}{2}-\lambda,\,1\right) \\ \left(\frac{\rho}{2},\,1\right),\,\left(\frac{1+\rho}{2},\,1\right),\,\left(\frac{\rho}{2}+\mu,\,1\right),\,\left(\frac{\rho}{2}-\mu,\,1\right) \end{matrix}\right].$$

$$z^{2\rho}\, W_{\lambda,\mu}(z)\, W_{-\lambda,\mu}(z) = 2^{2\rho}\,\pi^{-1/2} \times H^{4,0}_{2,4}\left[\frac{z^2}{4} \middle| \begin{matrix} (\rho+\lambda+1,\,1),\,(\rho-\lambda+1,\,1) \\ (\frac{1}{2}+\rho,\,1),\,(1+\rho,\,1),\,(\rho\pm\mu+\frac{1}{2},\,1) \end{matrix}\right].$$

$$z^{\rho}\, M_{\lambda,\mu}(iz)\, M_{\lambda,\mu}(-iz) = \frac{2^{\rho+1}\,\Gamma(\frac{1}{2})\,\Gamma^2(2\mu+1)}{\Gamma(\frac{1}{2}+\lambda+\mu)\,\Gamma(\frac{1}{2}-\lambda-\mu)} \times H^{1,2}_{2,4}\left[\frac{z^2}{4} \middle| \begin{matrix} \left(\frac{\rho}{2}+\lambda+1,\,1\right),\,\left(\frac{\rho}{2}+1-\lambda,\,1\right) \\ \left(\frac{\rho+1}{2}+\mu,\,1\right),\,\left(\frac{\rho+1}{2}-\mu,\,1\right),\,\left(\frac{\rho}{2}+1,\,1\right),\,\left(\frac{\rho+1}{2},\,1\right) \end{matrix}\right].$$

$$z^{\lambda}\, J_{\mu}(z)\, J_{\nu}(z) = \frac{1}{\Gamma(\frac{1}{2})} \times H^{1,2}_{2,4}\left[z^2 \middle| \begin{matrix} \left(\frac{\lambda}{2},\,1\right),\,\left(\frac{\lambda+1}{2},\,1\right) \\ \left(\frac{\lambda+\mu+\nu}{2},\,1\right),\left(\frac{\lambda-\mu+\nu}{2},\,1\right),\left(\frac{\lambda+\mu-\nu}{2},\,1\right),\left(\frac{\lambda-\mu-\nu}{2},\,1\right) \end{matrix}\right].$$

$$z^{\lambda} J_{\nu}^{2}(z) = \frac{1}{\Gamma(\frac{1}{2})} H_{1,3}^{1,1}\left[z^2 \left| \begin{matrix} \left(\frac{\lambda+1}{2}, 1\right) \\ \left(\frac{\lambda}{2}+\nu, 1\right), \ \left(\frac{\lambda}{2}-\nu, 1\right), \ \left(\frac{\lambda}{2}, 1\right) \end{matrix} \right.\right].$$

$$z^{\lambda} J_{-\nu}(z) J_{\nu}(z) = \frac{1}{\Gamma(\frac{1}{2})} H_{1,3}^{1,1}\left[z^2 \left| \begin{matrix} \left(\frac{\lambda+1}{2}, 1\right) \\ \left(\frac{\lambda}{2}, 1\right), \left(\frac{\lambda}{2}+\nu, 1\right), \left(\frac{\lambda}{2}-\nu, 1\right) \end{matrix} \right.\right].$$

$$z^{\lambda} J_{\nu}(z) I_{\nu}(z) = 2^{3\lambda/2}\, \Gamma(\tfrac{1}{2})$$
$$\times H_{0,4}^{1,0}\left[\frac{z^4}{64} \left| \left(\frac{\lambda+2\nu}{4}, 1\right), \left(\frac{\lambda-2\nu}{4}, 1\right), \left(\frac{\lambda}{4}, 1\right), \left(\frac{\lambda+2}{4}, 1\right) \right.\right].$$

$$z^{\lambda} J_{\nu}(z)\, Y_{\nu}(z) = -\frac{1}{\Gamma(\frac{1}{2})} H_{1,3}^{2,0}\left[z^2 \left| \begin{matrix} \left(\frac{\lambda+1}{2}, 1\right) \\ \left(\frac{\lambda}{2}, 1\right), \left(\frac{\lambda}{2}+\nu, 1\right), \left(\frac{\lambda}{2}-\nu, 1\right) \end{matrix} \right.\right].$$

$$\left(\frac{z^2}{8}\right)^{\lambda} J_{-\nu}(z)\, I_{\nu}(z) = \Gamma\left(\frac{1}{2}\right)\cos\left(\frac{\nu\pi}{2}\right)$$
$$\times H_{0,4}^{1,0}\left[\frac{z^4}{64} \left| \left(\frac{\lambda}{2}, 1\right), \left(\frac{\lambda+1}{2}, 1\right), \left(\frac{\lambda+\nu}{2}, 1\right), \left(\frac{\lambda-\nu}{2}, 1\right) \right.\right]$$
$$-\Gamma\left(\frac{1}{2}\right)\sin\left(\frac{\nu\pi}{2}\right) H_{0,4}^{1,0}\left[\frac{z^4}{64} \left| \left(\frac{\lambda+1}{2}, 1\right), \left(\frac{\lambda}{2}, 1\right), \left(\frac{\lambda+\nu}{2}, 1\right), \left(\frac{\lambda-\nu}{2}, 1\right) \right.\right].$$

$$z^{2\lambda} I_{\nu}(z) K_{\nu}(z) = (4\pi)^{-1/2}\; H_{1,3}^{2,1}\left[z^2 \left| \begin{matrix} (\frac{1}{2}+\lambda, 1) \\ (\lambda+\nu, 1), (\lambda, 1), (\lambda-\nu, 1) \end{matrix} \right.\right].$$

$$z^{\lambda} I_{\nu}(z)\, K_{\mu}(z) = (4\pi)^{-1/2}$$
$$\times H_{2,4}^{2,2}\left[z^2 \left| \begin{matrix} \left(\frac{\lambda}{2}, 1\right), \left(\frac{\lambda+1}{2}, 1\right) \\ \left(\frac{\lambda+\mu+\nu}{2}, 1\right), \left(\frac{\lambda-\mu+\nu}{2}, 1\right), \left(\frac{\lambda+\mu-\nu}{2}, 1\right), \left(\frac{\lambda-\mu-\nu}{2}, 1\right) \end{matrix} \right.\right].$$

$$z^{\lambda} K_{\nu}(z)\, J_{\nu}(z) = \frac{1}{\Gamma(\frac{1}{2})} 2^{(3\lambda-1)/2}$$
$$\times H_{0,4}^{3,0}\left[\frac{z^4}{64} \left| \left(\frac{\lambda}{4}+\frac{\nu}{2}, 1\right), \left(\frac{\lambda}{4}+\frac{1}{2}, 1\right), \left(\frac{\lambda}{4}, 1\right), \left(\frac{\lambda}{4}-\frac{\nu}{2}, 1\right) \right.\right].$$

$$z^{\lambda} K_{\nu}(z) K_{\mu}(z) = \tfrac{1}{2}\,\Gamma(\tfrac{1}{2})$$

$$\times H^{4,0}_{2,4}\left[z^2 \middle| \begin{matrix} \left(\frac{\lambda}{2}, 1\right), \left(\frac{\lambda+1}{2}, 1\right) \\ \left(\frac{\lambda+\mu+\nu}{2}, 1\right), \left(\frac{\lambda-\mu+\nu}{2}, 1\right), \left(\frac{\lambda+\mu-\nu}{2}, 1\right), \left(\frac{\lambda-\mu-\nu}{2}, 1\right) \end{matrix} \right].$$

$$z^{\lambda} K^2_{\nu}(z) = \frac{1}{2}\,\Gamma\left(\frac{1}{2}\right) H^{3,0}_{1,3}\left[z^2 \middle| \begin{matrix} \left(\frac{\lambda+1}{2}, 1\right) \\ \left(\nu+\frac{\lambda}{2}, 1\right), \left(\frac{\lambda}{2}-\nu, 1\right), \left(\frac{\lambda}{2}, 1\right) \end{matrix} \right].$$

$$z^{\lambda} H^{(1)}_{\nu}(z)\, H^{(2)}_{\nu}(z) = \frac{2}{\pi^{5/2}} \cos(\nu\pi)$$

$$\times H^{3,1}_{1,3}\left[z^2 \middle| \begin{matrix} \left(\frac{\lambda+1}{2}, 1\right) \\ \left(\frac{\lambda}{2}+\nu, 1\right), \left(\frac{\lambda}{2}-\nu, 1\right), \left(\frac{\lambda}{2}, 1\right) \end{matrix} \right].$$

$$z^{2\lambda} K_{2\nu}(ze^{i\pi/4})\, K_{2\nu}(ze^{-i\pi/4})$$

$$= \frac{2^{3\lambda-3}}{\Gamma(\frac{1}{2})} H^{4,0}_{0,4}\left[\frac{z^4}{64} \middle| \left(\frac{\lambda}{2}, 1\right), \left(\frac{\lambda+1}{2}, 1\right), \left(\frac{\lambda}{2}+\nu, 1\right), \left(\frac{\lambda}{2}-\nu, 1\right) \right].$$

$$z^{\lambda}\,[J_{\nu}(z) J_{\mu}(z) + J_{-\nu}(z) J_{-\mu}(z)]$$

$$= \frac{2\cos\frac{\pi(\mu+\nu)}{2}}{\Gamma\left(\frac{1}{2}\right)}$$

$$H^{2,1}_{2,4}\left[z^2 \middle| \begin{matrix} \left(\frac{\lambda+1}{2}, 1\right), \left(\frac{\lambda}{2}, 1\right) \\ \left(\frac{\lambda+\mu+\nu}{2}, 1\right), \left(\frac{\lambda-\mu-\nu}{2}, 1\right), \left(\frac{\lambda-\mu+\nu}{2}, 1\right), \left(\frac{\lambda+\mu-\nu}{2}, 1\right) \end{matrix} \right].$$

$$z^{\lambda}\,[J_{\nu}(z) J_{\mu}(z) - J_{-\nu}(z) J_{-\mu}(z)]$$

$$= -\frac{2\sin\frac{\pi(\mu+\nu)}{2}}{\Gamma\left(\frac{1}{2}\right)}$$

$$H^{2,1}_{2,4}\left[z^2 \middle| \begin{matrix} \left(\frac{\lambda}{2}, 1\right), \left(\frac{\lambda+1}{2}, 1\right) \\ \left(\frac{\lambda+\mu+\nu}{2}, 1\right), \left(\frac{\lambda-\mu-\nu}{2}, 1\right), \left(\frac{\lambda-\mu+\nu}{2}, 1\right), \left(\frac{\lambda+\mu-\nu}{2}, 1\right) \end{matrix} \right]$$

$$z^{\lambda}\,[H_{\nu}^{(1)}(z)\,H_{\mu}^{(1)}(z)-H_{\nu}^{(2)}(z)\,H_{\mu}^{(2)}(z)]$$

$$=-4i\pi^{-1/2}$$

$$H_{2,4}^{3,0}\left[z^2\,\middle|\,\begin{matrix}\left(\frac{\lambda}{2},1\right),\left(\frac{\lambda+1}{2},1\right)\\ \left(\frac{\lambda+\mu+\nu}{2},1\right),\left(\frac{\lambda-\mu+\nu}{2},1\right),\left(\frac{\lambda+\mu-}{2},1\right),\left(\frac{\lambda-\mu-\nu}{2},1\right)\end{matrix}\right].$$

$$\frac{z^{\lambda}}{2i}\left[\exp\left\{\frac{i\pi}{2}(\nu-\mu)\right\}H_{\nu}^{(1)}(z)\,H_{\mu}^{(2)}(z)-\exp\left\{\frac{i\pi}{2}(\mu-\nu)\right\}H_{\mu}^{(1)}(z)\,H_{\nu}^{(2)}(z)\right]$$

$$=\pi^{-5/2}\{\cos(\mu\pi)-\cos(\nu\pi)\}$$

$$H_{2,4}^{4,1}\left[z^2\,\middle|\,\begin{matrix}\left(\frac{\lambda}{2},1\right),\left(\frac{\lambda+1}{2},1\right)\\ \left(\frac{\lambda+\mu+\nu}{2},1\right),\left(\frac{\lambda-\mu+\nu}{2},1\right),\left(\frac{\lambda+\mu-\nu}{2},1\right),\left(\frac{\lambda-\mu-\nu}{2},1\right)\end{matrix}\right].$$

$$\frac{z^{\lambda}}{2}\left[\exp\left\{\frac{i\pi}{2}(\nu-\mu)\right\}H_{\nu}^{(1)}(z)\,H_{\mu}^{(2)}(z)+\exp\left\{\frac{i\pi}{2}(\mu-\nu)\right\}H_{\mu}^{(1)}(z)\,H_{\nu}^{(2)}(z)\right]$$

$$=\pi^{-5/2}\{\cos(\mu\pi)+\cos(\nu\pi)\}$$

$$H_{2,4}^{4,1}\left[z^2\,\middle|\,\begin{matrix}\left(\frac{\lambda}{2},1\right),\left(\frac{\lambda+1}{2},1\right)\\ \left(\frac{\lambda+\mu+\nu}{2},1\right),\left(\frac{\lambda-\mu+\nu}{2},1\right),\left(\frac{\lambda+\mu-}{2},1\right),\left(\frac{\lambda-\mu-\nu}{2},1\right)\end{matrix}\right].$$

$$z^{\lambda}\,[I_{\nu}(z)\,I_{\mu}(z)-I_{-\nu}(z)\,I_{-\mu}(z)]$$

$$=-\frac{\sin(\mu+\nu)\,\pi}{\pi^{3/2}}$$

$$H_{2,4}^{2,0}\left[z^2\,\middle|\,\begin{matrix}\left(\frac{\lambda}{2},1\right),\left(\frac{\lambda+1}{2},1\right)\\ \left(\frac{\lambda+\mu+\nu}{2},1\right),\left(\frac{\lambda-\mu-\nu}{2},1\right),\left(\frac{\lambda-\mu+\nu}{2},1\right),\left(\frac{\lambda+\mu-\nu}{2},1\right)\end{matrix}\right].$$

$$z^{\lambda}\,{}_2F_1(\alpha,\beta;\gamma;-z)=\frac{\Gamma(\gamma)}{\Gamma(\alpha)\,\Gamma(\beta)}\,H_{2,2}^{1,2}\left[z\,\middle|\,\begin{matrix}(1+\lambda-\alpha,1),(1+\lambda-\beta,1)\\(\lambda,1),(1+\lambda-\gamma,1)\end{matrix}\right].$$

$$z^{\lambda}\,{}_2F_1(\alpha,\beta;\gamma;1-z)=\frac{\Gamma(\gamma)}{\Gamma(\alpha)\,\Gamma(\beta)\,\Gamma(\gamma-\alpha)\,\Gamma(\gamma-\beta)}$$

$$\times H_{2,2}^{2,2}\left[z\,\middle|\,\begin{matrix}(1+\lambda-\alpha,1),(1+\lambda-\beta,1)\\(\lambda,1),(\gamma-\alpha-\beta+\lambda,1)\end{matrix}\right],$$

where

$$\gamma-\alpha,\ \gamma-\beta \neq 0, -1, -2, \ldots .$$

$$z^{\lambda}\ {}_pF_q(a_p;\ b_q;\ -z)$$

$$= \frac{\prod_{j=1}^{q} \Gamma(b_j)}{\prod_{j=1}^{p} \Gamma(a_j)} H^{1,p}_{p,q+1}\left[z \left| \begin{matrix} (1+\lambda-(a_p), 1) \\ (\lambda, 1), (1+\lambda-(b_q), 1) \end{matrix} \right. \right],$$

$$p \leqslant q \quad \text{or} \quad p=q+1 \quad \text{and} \quad |z| < 1.$$

A8 HYPERGEOMETRIC FUNCTIONS OF TWO VARIABLES

APPELL'S FUNCTIONS

$$F_1(a, b, b'; c; x, y)$$

$$= \sum_{m=0}^{\infty} \sum_{n=0}^{\infty} \frac{(a)_{m+n}\,(b)_m\,(b')_n}{(c)_{m+n}\, m!\, n!} x^m y^n$$

$$= \sum_{m=0}^{\infty} \frac{(a)_m\,(b)_m}{(c)_m\, m!}\ {}_2F_1(a+m, b'; c+m, y)\, x^m,$$

where

$$x| < 1, |y| < 1.$$

$$F_2(a, b, b'; c, c'; x, y)$$

$$= \sum_{m=0}^{\infty} \sum_{n=0}^{\infty} \frac{(a)_{m+n}\,(b)_m\,(b')_n}{(c)_m\,(c')_n\, m!\, n!} x^m y^n$$

$$= \sum_{m=0}^{\infty} \frac{(a)_m\,(b)_m}{(c)_m\, m!}\ {}_2F_1(a+m, b', c'; y)\, x^m$$

where

$$|x| + |y| < 1.$$

$$F_3(a, a', b, b'; c; x, y)$$

$$= \sum_{m=0}^{\infty} \sum_{n=0}^{\infty} \frac{(a)_m\,(a')_n\,(b_m)\,(b')_n}{(c)_{m+n}\, m!\, n!} x^m y^n$$

$$= \sum_{m=0}^{\infty} \frac{(a)_m\,(b)_m}{(c)_m\, m!}\ {}_2F_1(a', b'; c+m; y)\, x^m,$$

where

$$|x| < 1 \quad \text{and} \quad |y| < 1.$$

$$F_4(a, b; c, c'; x, y)$$

$$= \sum_{m=0}^{\infty} \sum_{n=0}^{\infty} \frac{(a)_{m+n}\,(b)_{m+n}}{(c)_m\,(c')_n\, m!\, n!} x^m y^n$$

$$= \sum_{m=0}^{\infty} \frac{(a)_m\,(b)_m}{(c)_m\, m!}\ {}_2F_1(a+m, b+m; c'; y)\, x^m,$$

where

$$|x^{1/2}|+|y^{1/2}|<1.$$

Particular cases of Appell's functions

$$F_1(a, b, b'; a; x, y) = (1-x)^{-b}(1-y)^{-b'}.$$
$$F_2(a, b, b'; b, b'; x, y) = (1-x-y)^{-a}.$$
$$F_2(a, b, b'; a, b'; x, y) = (1-x-y)^{-b}(1-y)^{b-a}.$$
$$F_2(a, b, b'; b, a; x, y) = (1-x)^{b'-a}(1-x-y)^{-b'}.$$
$$xF_2(1, 1, b'; 2, b'; x, y)+yF_2(1, b, 1; b, 2; x, y) = -\log(1-x-y).$$

Confluent Hypergeometric Function of Two Variables

$$\Phi_1(a, b; c; x, y) = \sum_{m=0}^{\infty}\sum_{n=0}^{\infty}\frac{(a)_{m+n}(b)_m}{(c)_{m+n}\, m!\, n!}x^m y^n, \ |x|<1.$$

$$\Phi_2(b, b'; c; x, y) = \sum_{m=0}^{\infty}\sum_{n=0}^{\infty}\frac{(b_m)(b')_m}{(c)_{m+n}\, m!\, n!}x^m y^n$$

$$\Phi_3(b; c; x, y) = \sum_{m=0}^{\infty}\sum_{n=0}^{\infty}\frac{(b)_m}{(c)_{m+n}\, m!\, n!}x^m y^n.$$

$$\psi_1(a, b; c, c'; x, y) = \sum_{m=0}^{\infty}\sum_{n=0}^{\infty}\frac{(a)_{m+n}(b)_m}{(c)_m(c')_n\, m!\, n!}x^m y^n, \ |x|<1.$$

$$\psi_2(a; c, c'; x, y) = \sum_{m=0}^{\infty}\sum_{n=0}^{\infty}\frac{(a)_{m+n}}{(c)_m(c')_n\, m!\, n!}x^m y^n.$$

$$\Xi_1(a, a', b; c; x, y) = \sum_{m=0}^{\infty}\sum_{n=0}^{\infty}\frac{(a)_m(a')_n(b)_m}{(c)_{m+n}\, m!\, n!}x^m y^n, \ |x|<1.$$

$$\Xi_2(a, b; c; x, y) = \sum_{m=0}^{\infty}\sum_{n=0}^{\infty}\frac{(a)_m(b)_m}{(c)_{m+n}\, m!\, n!}x^m y^n, \ |x|<1.$$

Kampé de Fériet's Hypergeometric Function

$$F\left(\begin{array}{c|c|c} \mu & a_1, \ldots, a_\mu & \\ \nu & b_1, b_1', \ldots, b_\nu, b_\nu' & \\ & & x, y \\ \rho & c_1, \ldots, c_\rho & \\ \sigma & d_1, d_1', \ldots, d_\sigma, d_\sigma' & \end{array}\right)$$

$$= \sum_{m=0}^{\infty} \sum_{n=0}^{\infty} \frac{\prod_{j=1}^{\mu} (a_j)_{m+n} \prod_{j=1}^{\nu} \{(b_j)_m (b_j')_n\} x^m y^n}{\prod_{j=1}^{\rho} (c_j)_{m+n} \prod_{j=1}^{\sigma} \{(d_j)_m (d_j')_n\} m!\, n!}$$

$(\mu + \nu < \rho + \sigma + 1$ or $\mu + \nu = \rho + \sigma + 1$ and $|x| + |y| < \min(1, 2^{\rho-\mu+1})$;

$$= -\frac{1}{4\pi^2 K} \int_{-i\infty}^{+i\infty} \int_{-i\infty}^{+i\infty} \psi(s, t)\, \Gamma(-s)\, \Gamma(-t)\, (-x)^s (-y)^t \, ds\, dt$$

where

$$K = \frac{\prod_{j=1}^{\mu} \Gamma(a_j) \prod_{j=1}^{\nu} \{\Gamma(b_j)\, \Gamma(b_j')\}}{\prod_{j=1}^{\rho} \Gamma(c_j) \prod_{j=1}^{\sigma} \{\Gamma(d_j)\, \Gamma(d_j')\}}$$

and

$$\psi(s, t) = \frac{\prod_{j=1}^{\mu} \Gamma(a_j + s + t) \prod_{j=1}^{\nu} \Gamma(b_j + s)\, \Gamma(b_j' + t)}{\prod_{j=1}^{\rho} \Gamma(c_j + s + t) \prod_{j=1}^{\sigma} \Gamma(d_j + s)\, \Gamma(d_j' + t)}.$$

Interesting particular cases

$$F\left(\begin{array}{c|c|c} \mu & a_1, \ldots, a_\mu & \\ 0 & \ldots\ldots & x, y \\ \rho & c_1, \ldots, c_\rho & \\ 0 & \ldots\ldots & \end{array}\right) = {}_\mu F_\rho\left(\begin{array}{c} a_1, \ldots, a_\mu \\ c_1, \ldots\, c_\rho \end{array}; x + y\right).$$

$$F\left(\begin{array}{c|c|c} 0 & \ldots\ldots & \\ \nu & b_1, b_1', \ldots, b_\nu, b_\nu' & \\ 0 & \ldots\ldots & x, y \\ \sigma & d_1, d_1', \ldots, d_\sigma, d_\sigma' & \end{array}\right)$$

$$= {}_\nu F_\sigma\left(\begin{array}{c} b_1, \ldots, b_\nu \\ d_1, \ldots, d_\sigma \end{array}; x\right) {}_\nu F_\sigma\left(\begin{array}{c} b_1', \ldots, b_\nu' \\ d_1', \ldots, d_\sigma' \end{array}; y\right).$$

$$F\left(\begin{array}{c|c|c} \omega & a_1, \ldots, a_\omega & \\ 1 & b, b' & \\ \omega & c_1, \ldots, c_\omega & x, x \\ 0 & \ldots\ldots & \end{array}\right) = {}_{\omega+1}F_\omega\left(\begin{array}{c} a_1, \ldots, a_\omega, b + b' \\ c_1, \ldots, c_\omega \end{array}; x\right).$$

For further results on Appell functions and other hypergeometric functions of two variables discussed here, see the monograph by Appéll and Kampé de Fériet (1926).

A9 HYPERGEOMETRIC FUNCTIONS OF SEVERAL VARIABLES

Lauricella's Hypergeometric Function of n Variables

$$F_A(a; b_1, \ldots, b_n; c_1, \ldots, c_n; x_1, \ldots, x_n)$$

$$= \sum_{m_1=0}^{\infty} \cdots \sum_{m_n=0}^{\infty} \frac{(a)_{m_1+\ldots+m_n} (b_1)_{m_1} \ldots (b_n)_{m_n} \, x_1^{m_1} \ldots x_n^{m_n}}{(c_1)_{m_1} \ldots (c_n)_{m_n} (m_1)! \ldots (m_n)!},$$

where

$$|x_1| + |x_2| + \ldots + |x_n| < 1.$$

$$F_B(a_1, \ldots, a_n; b_1, \ldots, b_n; c; x_1, \ldots, x_n)$$

$$= \sum_{m_1=0}^{\infty} \cdots \sum_{m_n=0}^{\infty} \frac{(a_1)_{m_1} \ldots (a_n)_{m_n} (b_1)_{m_1} \ldots (b_n)_{m_n}}{(c)_{m_1+\ldots+m_n} (m_1)! \ldots (m_n)!} x_1^{m_1} \ldots x_n^{m_n},$$

where

$$|x_1| < 1, |x_2| < 1, \ldots, |x_n| < 1.$$

$$F_C(a, b; c_1, \ldots, c_n; x_1, \ldots, x_n)$$

$$= \sum_{m_1=0}^{\infty} \cdots \sum_{m_n=0}^{\infty} \frac{(a)_{m_1+\ldots+m_n} (b)_{m_1+\ldots+m_n}}{(c_1)_{m_1} \ldots (c_n)_{m_n} (m_1)! \ldots (m_n)!} x_1^{m_1} \ldots x_n^{m_n},$$

where

$$|\sqrt{x_1}| + |\sqrt{x_2}| + \ldots + |\sqrt{x_n}| < 1.$$

$$F_D(a; b_1, \ldots, b_n; c; x_1, \ldots, x_n)$$

$$= \sum_{m_1=0}^{\infty} \cdots \sum_{m_n=0}^{\infty} \frac{(a)_{m_1+\ldots+m_n} (b_1)_{m_1} \ldots (b_n)_{m_n}}{(c)_{m_1+\ldots+m_n} (m_1)! \ldots (m_n)!} x_1^{m_1} \ldots x_n^{m_n},$$

where

$$|x_1| < 1, |x_2| < 1, \ldots, |x_n| < 1.$$

A10 MULTIPLE INTEGRAL REPRESENTATIONS

$$F_A(a; b_1, \ldots, b_n; c_1, \ldots, c_n; x_1, \ldots, x_n)$$

$$= \frac{\Gamma(c_1) \ldots \Gamma(c_n)}{\Gamma(b_1) \ldots \Gamma(b_n) \Gamma(c_1 - b_1) \ldots \Gamma(c_n - b_n)}$$

$$\times \int_0^1 \cdots \int_0^1 u_1^{b_1-1} \ldots u_n^{b_n-1} (1-u_1)^{c_1-b_1-1} \ldots (1-u_n)^{c_n-b_n-1}$$

$$\times (1 - u_1 x_1 - \ldots - u_n x_n)^{-a} \, du_1 \ldots du_n;$$

$$F_B(a_1, \ldots, a_n; b_1, \ldots, b_n; c; x_1, \ldots, x_n)$$

$$= \frac{\Gamma(c)}{\Gamma(a_1)\ldots\Gamma(a_n)\,\Gamma(c - a_1 - \ldots - a_n)}$$

$$\times \int \ldots \int u_1^{a_1-1} \ldots u_n^{a_n-1} (1 - u_1 - \ldots - u_n)^{c-a_1-\ldots-a_n-1}$$

$$\times (1 - u_1 x_1)^{-b_1} \ldots (1 - u_n x_n)^{-b_n}\, du_1 \ldots du_n$$

$$(u_1 \geqslant 0, \ldots, u_n \geqslant 0,\ 1 - u_1 - \ldots - u_n \geqslant 0);$$

$$F_D(a; b_1, \ldots, b_n; c; x_1, \ldots, x_n)$$

$$= \frac{\Gamma(c)}{\Gamma(b_1)\ldots\Gamma(b_n)\,\Gamma(c - b_1 - \ldots - b_n)}$$

$$\times \int \ldots \int u_1^{b_1-1} \ldots u_n^{b_n-1} (1 - u_1 - \ldots - u_n)^{c-b_1-\ldots-b_n-1}$$

$$\times (1 - u_1 x_1 - \ldots - u_n x_n)^{-a}\, du_1 \ldots du_n$$

$$(u_1 \geqslant 0, \ldots, u_n \geqslant 0,\ 1 - u_1 - \ldots - u_n \geqslant 0).$$

A11 CASES OF REDUCIBILITY

Cases of Reducibility of F_D

$$F_D(a; b_1, \ldots, b_n; c; x, \ldots, x)$$

$$= {}_2F_1(a, b_1 + \ldots + b_n; c; x)$$

$$F_D(a, b_1, \ldots, b_n;\ c;\ 1, 1, \ldots, 1)$$

$$= \frac{\Gamma(c)\,\Gamma(c - a - b_1 - \ldots - b_n)}{\Gamma(c - a)\,\Gamma(c - b_1 - \ldots - b_n)}$$

A single integral representation for the function F_D is given by

$$F_D(a; b_1, \ldots, b_n; c; x_1, \ldots, x_n)$$

$$= \frac{\Gamma(c)}{\Gamma(a)\,\Gamma(c - a)} \int_0^1 u^{a-1}(1 - u)^{c-a-1}(1 - ux_1)^{-b_1} \ldots (1 - ux_n)^{-b_n}\, du,$$

where $R(a) > 0,\ R(c - a) > 0.$

Confluent Hypergeometric Functions of Several Variables

$$\phi_2(b_1, \ldots, b_n; c; x_1, \ldots, x_n)$$

$$= \sum_{m_1=0}^{\infty} \ldots \sum_{m_n=0}^{\infty} \frac{(b_1)_{m_1} \ldots (b_n)_{m_n}}{(c)_{m_1+\ldots+m_n}} \frac{x_1^{m_1} \ldots x_n^{m_n}}{(m_1)! \ldots (m_n)!}$$

$$\psi_2(a; c_1, \ldots, c_n; x_1, \ldots, x_n)$$

$$= \sum_{m_1=0}^{\infty} \ldots \sum_{m_n=0}^{\infty} \frac{(a)_{m_1+\ldots+m_n} x_1^{m_1} \ldots x_n^{m_n}}{(c_1)_{m_1} \ldots (c_n)_{m_n} m_1! \ldots m_n!}$$

$$M_{k,\mu_1,\ldots,\mu_n}(x_1, \ldots, x_n)$$

$$= x_1^{\mu_1+1/2} \ldots x_n^{\mu_n+1/2} \exp\left[-\frac{x_1 + \ldots + x_n}{2}\right]$$

$$\times \psi_2\left(\mu_1 + \ldots + \mu_n - k + \frac{n}{2}; 2\mu_1 + 1, \ldots, 2\mu_n + 1; x_1, \ldots, x_n\right).$$

For a detailed definition and other properties of these functions, see the original paper by Pierre Humbert [*La fonction* $W_{k,\mu_1,\ldots,\mu_n}(x_1, \ldots, x_n)$, *Comptes rendus*, t. CLXXI, 1920, p. 428] and Appéll and Kampé de Fériet (1926).

Bibliography

ABIODUN, R.F.A. AND SHARMA, B.L. (1971) (*Also see* SHARMA, B.L.). Summation of series involving generalized hypergeometric functions of two variables. *Glasnik Mat*. Ser. III 6 (26), 253–264.

ABIODUN, R.F.A. AND SHARMA, B.L. (1973). Fourier series for generalized function of two variables. *Univ. Nac. Tucumàn Rev*. Ser. A, 23, 25–33.

AGARWAL, B.M. (1968). Application of Δ and E operators to evaluate certain integrals. *Proc. Cambridge Philos. Soc.* 64, 99–104.

AGARWAL, B.M. (1968a). On generalized Meijer H-functions satisfying the Truesdell F-equations. *Proc. Nat. Acad. Sci.* India Sect. A, 38, 259–264.

AGARWAL, B.M. AND SINGHAL, B.M. (1974). A transformation from G-functions to H-functions. *Vijnana Parishad Anusandhan Patrika* 17, 137–142.

AGARWAL, I. AND SAXENA, R.K. (1969). Integrals involving Bessel functions. *Univ. Nac. Tucumàn Rev*. Ser. A, 19, 245–254.

AGARWAL, I. AND SAXENA, R.K. (1972). An infinite integral involving Meijer's G-function. *Riv. Mat. Univ. Parma* (3) 1, 15–21.

AGARWAL, R.P. (1965). An extension of Meijer's G-function. *Proc. Nat. Inst. Sci.* India Part A, 31, 536–546.

AGARWAL, R.P. (1970). On certain transformation formulae and Meijer's G-function of two variables. *Indian J. Pure Appl. Math.* 1, No. 4, 537–551.

AGARWAL, R.P. (1973). Contributions to the theory of generalized hypergeometric series. *J. Math. Phys. Sci.* Madras 7, Jubilaums—Sonderheft, S93–S100.

AGGARWALA, INDRA AND GOYAL, A.N. (1973). On some integrals involving generalized Lommel, Maitland and A*-functions. *Indian J. Pure. Appl. Math.* 4, 798–805.

AL-SALAM, W.A. (AL-SALAM, WALEED, A; *Also see* CARLITZ, L.) (1966–67). Some fractional q-integrals and q-derivatives. *Proc. Edinburgh Math. Soc.* (2) 15, 135–140.

ANANDANI, P. (1967). Some expansion formulae for H-function II. *Ganita* 18, 89–101.

ANANDANI, P. (1968). Some integrals involving products of Meijer's G-function and H-function. *Proc. Indian Acad. Sci.* Sect. A, 67, 312–321.

ANANDANI, P. (1968a). Summation of some series of products of H-functions. *Proc. Nat. Inst. Sci.* India Part A, 34, 216–223.

ANANDANI, P. (1968b). Fourier series for H-function. *Proc. Indian Acad. Sci.* Sect. A, 68, 291–295.

ANANDANI, P. (1969). On some integrals involving generalized associated Legendre's functions and H-functions. *Proc. Nat. Acad. Sci.* India Sect. A, 39, 341–348.

ANANDANI, P. (1969a). On some recurrence formulae for the H-function. *Ann. Polon. Math.* 21, 113–117.

ANANDANI, P. (1969b). On finite summation, recurrence relations and identities of H-functions. *Ann. Polon. Math,* 21, 125–137.

ANANDANI, P. (1969c). Some infinite series of H-function—I. *Math. Student* 37, 117–123.

ANANDANI, P. (1969d). Some integrals involving products of generalized Legendre's associated functions and the H-function. *J. Sci. Eng. Res.* 13, 274–279.

ANANDANI, P. (1969e), Some integrals involving generalized associated Legendre's functions and the H-function. *Proc. Nat. Acad. Sci.* India Sect. A, 39, 127–136.

ANANDANI, P. (1969f). Some expansion formulae for the H-function—III. *Proc. Nat. Acad. Sci.* India Sect. A, 39, 23–34.

ANANDANI, P. (1969g). Some expansion formulae for H-function. IV. *Rend. Circ. Mat. Palermo* (2) 18, 197–214.

ANANDANI, P. (1969h). On some identities of H-function. *Proc. Indian Acad. Sci.* Sect. A, 70, 89–91.

ANANDANI, P. (1969i). Some integrals involving H-functions. *Labdev. J. Sci. Tech.* Part A, 7, 62–66.

ANANDANI, P. (1970). Some integrals involving associated Legendre functions of the first kind and the H-function. *J. Natur. Sci. and Math.* 10, 97–104.

ANANDANI, P. (1970a). Some infinite series of H-functions. II. *Vijnana Parishad Anusandhan Patrika* 13, 57–66.

ANANDANI, P. (1970b). Integration of products of generalized Legendre function with respect to parameters. *Labdev J. Sci. Tech.* 9A, 13–19.

ANANDANI, P. (1970c). On the derivative of H-function. *Rev. Roum. Math. Pures et Appl.* 15, 189–191.

ANANDANI, P. (1970d). Use of generalized Legendre associated function and the H-function in heat production in a cylinder. *Kyungpook Math. J.* 10, 107–113.

ANANDANI, P. (1970e). Some integrals involving Jacobi polynomials and H-function. *Labdev. J. Sci. Tech.* Part A, 8, 145–149.

ANANDANI, P. (1970f). An expansion formula for the H-function involving associated Legendre function. *J. Natur. Sci. Math.* 10, No. 1, 49–51.

ANANDANI, P. (1970g). An expansion for the H-function involving generalized Legendre's associated functions. *Glasnik Mat.* Ser. III 5 (25), 55–58.

ANANDANI, P. (1970h). Expansion of the H-function involving generalized Legendre's associated function and H-function. *Kyungpook Math. J.* 10, 53–57.

ANANDANI, P. (1970i). Some expansion formulae for the H-function. *Labdev. J. Sci. Tech.* India Part A, 8, 80–87.

ANANDANI, P. (1970j). Some integrals involving H-functions of generalized arguments. *Math. Education* 4, 32–38.

ANANDANI, P. (1970k). Some infinite series of H-function II. *Vijnana Parishad Anusandhan Patrika* 13, 57–66.

ANANDANI, P. (1971). Some integrals involving associated Legendre functions and the H-function. *Univ. Nac. Tucumàn. Rev.* Ser. A, 21, 33–41.

ANANDANI, P. (1971a). An expansion formula for the H-function involving products of associated Legendre functions and H-functions. *Univ. Nac. Tucumàn Rev.* Ser. A, 21, 95–99.

ANANDANI, P. (1971b). Some integrals involvinng H-function. *Rend. Circ, Mat. Palermo* (2), 20, 70–82.

ANANDANI, P. (1971c). An expansion formula for the H-function involving generalized Legendre associated functions. *Portugal Math.* 30, 173–180.

ANANDANI, P. (1971d). Integration of products of generalized Legendre functions and the H-function with respect to parameters. *Labdev. J. Sci. Tech.* Part A, 9, 13–19.

ANANDANI, P. (1972). On some generating functions for the H-functions. *Labdev. J. Sci, Tech.* Part A, 10, 5–8.

ANANDANI, P. (1972a) Some infinite series of H-functions. *Ganita* 23, No. 2, 11–17.

ANANDANI, P. (1973). Integrals involving products of generalized Legendre functions and the H-function. *Kyungpook Math. J.* 13, 21–25.

ANANDANI, P. (1973a). Some integrals involving the H-function and generalized Legendre functions. *Bull. Soc. Math. Phys. Macedoine* 24, 33–38.

ANANDANI, P. (1973b). Expansion theorems for the H-function involving associated Legendre functions. *Bull. Soc. Math. Phys. Macedoine* 24, 39–43.

ANANDANI, P. (1973c). On some results involving generalized Legendre's associated functions and H-functions. *Ganita* 24, No. 1, 41–48.

ANANDANI, P. AND SRIVASTAVA, H.S.P. (1972/73). On Mellin transform of product involving Fox's H-function and a generalized function of two variables. *Comment Math. Univ. St. Paul*, 21, fasc. 2, 35–42.

ANDREWS, GEORGE. E. (1974), Applications of basic hypergeometric functions. *Siam Review* 16, 441–484.

APPELL, P. AND KAMPÉ DE FÉRIET, J. (1926). *Fonctions Hypergéometriques et Hypersphèriques; Polynomes d'Hermites.* Paris, Gauthier-Villars.

ASKEY, R. (1965). Orthogonal expansions with positive coefficients. *Proc. Amer. Math. Soc.* 16, 1191–1194.

BAILEY, W.N. (1933). A reducible case of the fourth type of Appell's hypergeometric function of two variables. *Quart. J. Math.* Oxford Ser. (2) 4, 305–308.

BAILEY, W.N. (1934). On the reducibility of Appell's function F_4. *Quart. J. Math.* Oxford Ser. (2) 5, 291–292.

BAILEY, W.N. (1935) *Generalized hypergeometric series.* Cambridge Tracts in Mathematics and Mathematical Physics No. 32, Cambridge University Press. Cambridge & New York.

BAILEY, W.N. (1936). Some integrals involving Bessel functions. *Proc. London. Math. Soc.* 40, 37–48.

BAILEY, W.N. (1936a). Some infinite integrals involving Bessel functions—II. *J. London Math. Soc.* 11, 16–20.

BAJPAI, S.D. (1969). An integral involving Fox's H-function and Whittaker functions. *Proc. Cambridge Philos. Soc.* 65, 709–712.

BAJPAI, S.D. (1969a). On some results involving Fox's H-function and Jacobi polynomials. *Proc. Cambridge Philos. Soc.* 65, 697–701.

BAJPAI, S.D. (1969b). Fourier series of generalized hypergeometric functions. *Proc. Cambridge Philos. Soc.* 65, 703–707.

BAJPAI, S.D. (1969c). An expansion formula for Fox's H-function. *Proc. Cambridge Philos. Soc.* 65, 683–685.

BAJPAI, S.D. (1969d). An integral involving Fox's H-function and heat conduction. *Math. Education* 3, 1–4.

BAJPAI, S.D. (1969e). An expansion formula for H-function involving Bessel functions. *Labdev. J. Sci. Tech.* Part A, 7, 18–20.

BAJPAI, S.D. (1969/70). An integral involving Fox's H-function and its application. *Univ. Lisboa, Revista Fac. ci.*, II. Ser. A, 13, 109–114.

BAJPAI, S.D. (1970). Some expansion formulae for Fox's H-function involving exponential functions. *Proc. Cambridge Philos. Soc.* 67, 87–92.

BAJPAI, S.D. (1970a). On some results involving Fox's H-function and Bessel function. *Proc. Indian Acad. Sci.* Sect. A, 72, 42–46.

BAJPAI, S.D. (1970b). Transformation of an infinite series of Fox's H-function. *Portugal Math.* 29, 141–144.

BAJPAI, S.D. (1971). Some results involving Fox's H-function. *Portugal Math.* 30, 45–52.

BAJPAI, S.D. (1972). Some results involving G-function of two variables. *Gaz. Mat. Lisboa* 33, 13–24.

BAJPAI, S.D, (1974). Expansion formulae for the products of Meijer's G-function and Bessel functions. *Portugal Math.* 33, 35–41.

BANERJI, P.K. AND SAXENA, R.K. (1971). Integrals involving Fox's H-function. *Bull. Math. Soc. Sci. Math.* R.S. Roumanie 15 (63), 263–269.

BANERJI, P.K. AND SAXENA, R.K. (1973). On some results involving products of H-functions. *An Sti. Univ. "Al. I. cuza"*, Iasi Sect. Ia Mat. (N.S.) 19, 175–178.

BANERJI, P.K. AND SAXENA, R.K. (1973a). Contour integral involving Legendre polynomial and Fox's H-function. *Univ. Nac. Tucuman Rev.* Ser. A, 23, 193–198.

BANERJI, P.K. AND SAXENA, R.K. (1976). Expansions of generalized H-functions. *Indian J. Pure and Appl. Math.* 7, No. 3, 337–341.

BARNES, E.W. (1908). A new development of the theory of the hypergeometric functions. *Proc. London Math. Soc.* (2) 6, 141–177.

BHAGCHANDANI, L.K. AND MEHRA, K.N. (1970). Some results involving generalized Meijer function and Jacobi polynomials. *Univ. Nac. Tucumàn Rev.* Ser. A, 20, 167–174.

BHATNAGAR, P.L. (1973). Numerical integration of Lommel type of integrals involving products of three Bessel functions. *Indian J. Math.* 15, 77–97.

BHATT, R.C. (1966). Certain integrals involving the products of hypergeometric functions. *Mathematische (Catania)* 21, 6–10.

BHISE, V.M. (1964). Some finite and infinite series of Meijer-Laplace transform. *Math. Ann.* 154, 267–272.

BHISE, V.M. (1967). Certain properties of Meijer-Laplace transform. *Comp. Math.* 18, 1–6.

BHONSLE, B.R. (1962). Some series and recurrence relations for MacRobert's E-function. *Proc. Glasgow Math. Assoc.* 5, 116–117.

BHONSLE, B.R. (1966). Jacobi polynomials and heat production in a cylinder. *Math. Japon.* 11, No. 1, 83–90.

BHONSLE, B.R. (1967). Steady state heat flow in a shell enclosed between two prolate spheroids. *Math. Japon.* 12 (1), 83–90.

BOCHNER, S. (1952). Bessel functions and modular relations of higher type and hyperbolic differential equations. *Com. Sem. Math. de l'Univ. de Lund*, Tome Supplementaire dediè á Marcel Riez, 12–20.

BOCHNER, S. (1958). On Riemann's functional equation with multiple gamma factors. *Ann. Math.* (2) 67, 29–41.

BOERSMA, J. (1962). On a function which is a special case of Meijer's G-function. *Comp. Math.* 15, 34–63.

BORA, S.L. (1970). An infinite integral involving generalized function of two variables. *Vijnana Parishad Anusandhan Patrika* 13, 95–100.

BORA, S.L. AND KALLA, S.L. (1970). Some results involving generalized function of two variables. *Kyungpook Math. J.* 10, 133–140.

BORA, S.L. AND KALLA, S.L. (1971). Some recurrence relations for the H-function. *Vijnana Parishad Anusandhan Patrika* 14, 9–12.

BORA, S L. AND KALLA, S.L. (1971a). An expansion formula for the generalized function of two variables. *Univ. Nac. Tucumàn Rev.* Ser. A, 21, 53–58.

BORA, S.L., KALLA, S.L. AND SAXENA, R.K. (1970). On integral transforms. *Univ. Nac. Tucumàn Rev.* Ser. A, 20, 181–188.

BORA, S.L., AND SAXENA, R.K. (1971). Integrals involving product of Bessel functions and generalized hypergeometric functions. *Pub. Inst. Math.* (Beograd) 11(25), 23–28.

BORA, S.L. SAXENA, R.K. AND KALLA, S.L. (1972), An expansion formula for Fox's H-function of two variables. *Univ. Nac. Tucumàn Rev*. Ser. A, 22, 43–48.

BRAAKSMA, B.L.J. (1964). Asymptotic expansions and analytic continuations for a class of Barnes integrals. *Comp. Math*. 15, 239–341.

BROMWICH, T.J.I'A (1909). An asymptotic formula for the generalized hypergeometric series. *Proc. London. Math. Soc*. (2) 7, 101–106.

BURCHNALL, J.L. (1939). The differential equations of Appell's function F_4. *Quart. J. Math*. Oxford 10, 145–150.

BURCHNALL, J.L. (1942). Differential equations associated with hypergeometric functions. *Quart. J. Math*. Oxford Ser. 13, 90–106.

BURCHNALL, J.L. AND CHAUNDY, T.W. (1940). Expansions of Appell's double hypergeometric functions. *Quart J. Math*. Oxford Ser. 11, 249–270.

BURCHNALL, J.L. AND CHAUNDY, T.W. (1941). Expansions of Appell's double hypergeometric functions—II. *Quart J. Math*. Oxford Ser. 12, 112–128.

BUSCHMAN, R.G. (1972). Contiguous relations and related formulas for the H-function of Fox. *Jñānabha* Sect. A, 2, 39–47.

BUSCHMAN, R.G. (1974). The asymptotic expansion of an integral. *Rendiconti di Matematica* (3) 7, Series 6, 481–486.

BUSCHMAN, R.G. (1974a). Partial derivatives of the H-function with respect to parameters expressed as finite sums and as integrals. *Univ. Nac. Tucumàn Rev*. Ser. A, 24, 149–155.

BUSCHMAN, R.G. (1974b). Finite sum representations for partial derivatives of special functions with respect to parameters. *Math. Comp*., 28, No. 127, 817–824.

BUSCHMAN, R.G. AND GUPTA, K.C. (1975). Contiguous relations for the H-functions of two variables. *Indian J. Pure Appl. Math*. 6, No. 12, 1416–1421.

BUSCHMAN, R.G. AND SRIVASTAVA, H.M. (1975) Inversion formulas for the integral transformation with the H-function as kernel. *Indian J. Pure Appl. Math*. 6, No. 6, 583–590.

CARLITZ, L. (1962) Summation of some series of Bessel functions. *Nederl. Akad. Wetensch. Proc*. Ser. A, 65=Indag. Math. 24, 47–54.

CARLITZ, L. AND AL-SALAM, W.A. (1963). Some functions associated with Bessel functions. *J. Math. Mech*. 12, 911–933.

CARLSON, B.C. (1963). Lauricella's hypergeometric function F_D. *J. Math. Anal. Appl*. 7, 452–470.

CHAK, A.M. (1970). Some generalization of Laguerre polynomials, I, II. *Math. Vesnik* 7 (22), 7–13, 14–18.

CHANDEL, R.C.S. (1969). Generalized Laguerre polynomials and the polynomials related to them. *Indian J. Math*. 11, 57–66.

CHANDEL R.C.S. (1971). A short note on generalized Laguerre polynomials and the polynomials related to them. *Indian J. Math*. 13, 25–27.

CHANDEL, R.C.S. (1972). Generalized Laguerre polynomials and the polynomials related to them II. *Indian J. Math*. 14, 149–155.

CHANDEL, R.C.S. (1973). On some multiple hypergeometric functions related to Lauricella functions. *Jñānabha* (A) 3, 119–136.

CHANDEL, R.C.S. AND AGARWAL, R.D. (1971). On the G-functions of two variables. *Jñānabha* Sect. A. 1, No. 1, 83–91.

CHHABRA, S.P. AND SINGH, F. (1969). An integral involving product of a G-function and a generalized hypergeometric function. *Proc. Cambridge Philos. Soc*. 65, 479–482.

CHANDRASEKHARAN, K. AND NARASIMHAN RAGHAVAN (1962). Functional equations with multiple gamma factors and the average order of arithmetical functions. *Ann. Math*. 76, 93–136.

CHATTERJEA, S.K. (1964). On a generalization of Laguerre polynomials. *Rend Sem. Math*. Univ. Padova 34, 180–190.

CHATURVEDI, K.K. AND GOYAL, A.N. (1972). A*-function, I. *Indian J. Pure Appl. Math.* 3, 357–360.

CHURCHILL, R.V. (1941). *Fourier series and boundary value problems.* McGraw Hill, New York.

CONSTANTINE, A.G. (1963). Some non-central distribution problems in multivariate analysis. *Ann. Math. Statit.* 34, 1270–1285.

CONSTANTINE, A.G. AND MUIRHEAD, R.J. (1972). Partial differential equations of hypergeometric functions of two arguments matrices. *J. Multivariate Analysis* 2. 332–338.

DAHIYA, R.S. (1971). Multiple integrals and the transformations involving H-functions and Tchebichef polynomials. *Acta Mexicana Ci. Tech.* 5, 192–197.

DAHIYA, R.S. (1971a). On integral representation of Fox's H-function for evaluating double integrals. *An. Fac. Ci.* Univ. Porto 54, 363–367.

DAHIYA, R.S. (1971/72). On an integral relation involving Fox's H-function. *Univ. Lisboa Revista Fac. Ci. A* (2) 14, 105–111.

DAHIYA, R.S. AND SINGH, B. (1971). Fourier series of Meijer's G-function of higher order. *An. Sti. Univ. Al. I. Cuza. n. Ser.* Sect. (1) 17, 111–116.

DAHIYA, R.S. AND SINGH, B (1972). On Fox's H-function. *Indian J. Pure Appl. Math.* 3. No. 3, 493–495.

DE AMIN, L.H. AND KALLA, S.L. (1973). Integrales que involucran productos de funciones. hipergeometricas generalizadas y la function H de dos variables. *Univ. Nac. Tucumàn Rev.* Ser. A, 23, 131–141.

D'ANGELO, IDLA. G. AND KALLA, S.L. (1973). Algunos resultados que involucran la function H de Fox. *Univ. Nac. Tucumàn Rev.* Ser. A, 33, 83–87.

DE ANGUIO, M.E.F. AND KALLA, S.L. (1972). The Laplace transform of the product of two Fox's H-functions. *Univ. Nac. Tucumàn Rev.* Ser. A, 22, 171–175.

DE ANGUIO, M.E.F., DE GOMEZ LOPEZ, A.M.M. AND KALLA, S.L. (1972). Integrals that involve the H-function of two variables. *Acta Mexicana Ci. Tecn.* 6, 30–41.

DE ANGUIO, M.E.F., DE GOMEZ LOPEZ, A.M.M. AND KALLA, S.L. (1972a). Integrales que involucrana la function H de dos variables. *Acta Mexicana Cien. Tecn.* (2) 6, 30–41.

DE ANGUIO, M.E.F. AND KALLA, S.L. (1973). Sobre Integracion con respecto a parametros. *Univ. Nac. Tucumàn Rev.* Ser. A, 23, 103–110.

DE BATTIG, N.E.F. AND KALLA, S.L. (1971). Some results involving generalized hypergeometric function of two variables. *Rev. Ci.* Mat. Univ. Laurenco Marques Ser. A, 2, 47–53.

DE BATTIG, N.E.F. AND KALLA, S.L. (1971a). On certain finite integrals involving the hypergeometric, H-function of two variables. *Acta Mexicana Ci. Tecn.* 5, 142–148.

DE GALINDO, SANTANA, M. AND KALLA S.L. (1975). Sobre una extension de la function generalizada de dos variables. *Univ. Nac. Tucumàn. Rev.* Ser. A, 25, 221–229.

DE GOMEZ, LOPEZ, A.M.M. AND KALLA, S.L. (1972). Integrals that involve Fox's H-function. *Univ. Nac. Tucumàn Rev.* Ser. A, 22, 165–170.

DE GOMEZ, LOPEZ, A.M.M. AND KALLA, S.L. (1973). On a generalized integral transform. *Kyungpook Math. J.* 13 (2), 275–280.

DENIS, R.Y. (1968). Certain transformations of bilateral cognate trigonometrical series of hypergeometric type. *Proc. Cambridge Philos. Soc.* 64, 421–424.

DENIS, R.Y. (1969). A general expansion theorem for products of generalized hypergeometric series. *Proc. Nat. Inst. Sci.* India, Part A, 35, 70–76.

DENIS, R.Y. (1970) Certain integrals involving G-function of two variables. *Ganita* 21, No. 2, 1–10.

DENIS, R.Y. (1972). Certain expansions of generalized hypergeometric series. *Math. Student* 40A, 82–86.

DENIS, R.Y. (1973). On certain double series involving generalized hypergeometric series. *Bull. Soc. Math. Phys. Macédoine* 23, 33–35.

DESHPANDE, V.L. (1971). Expansion theorems for the Kampé de Fériet function. *Nederl. Akad. Wetensch. Proc.* 74=Indag. Math. 33, 39–45.

DESHPANDE, V.L. (1971a). On the derivatives of G-function of two variables. *Proc. Nat. Acad. Sci.* India Sect. A, 41, 60–68.

DHAWAN, G.K. (1969). Series and expansion formulae for G-function of two variables. *J. M.A.C.T.* 2, 88–94.

DIXON, A.L. AND FERRAR, W.L. (1936). A class of discontinuous integrals. *Quart. J. Math.* Oxford Ser. 7, 81–96.

DOETSCH, G. (1943). *Theorie and Anwendung der Laplace-Transformation.* Dover, New York.

DOETSCH, G. (1958). *Einfuhrung in Theorie und Anwendung der Laplace-Transformation.* Birkhäuser-Verlag, Basel.

DUBEY, G.K. AND SHARMA, C.K. (1972). On Fourier series for generalized Fox H-functions. *Math. Student* 40A (1972), 147–156.

DZRBASJAN, V.A. (1964). On a theorem of Whipple Z. Wyčysl. *Mat. i. Mat. Fiz,* 4, 348–351.

EDELSTIEN L.A. (1964). On the one-centre expansion of Meijer's G-function. *Proc. Cambridge Philos. Soc.* 60, 533–538.

ERDÉLYI, A. (1950–51). On some functional transformations. *Univ. Politec. Torino Rend. Sem. Mat.* 10, 217–234.

ERDÉLYI, A. (1954). On a generalization of the Laplace-Transformation. *Proc. Edinburgh Math. Soc.* (2) 10, 53–55.

ERDÉLYI, A., MAGNUS, W., OBERHETTINGER, F. AND TRICOMI, F.G. (1953). *Higher transcendental functions,* Vol. I. II, McGraw-Hill, New York.

ERDÉLYI, A., MAGNUS, W., OBERHETTINGER, F. AND TRICOMI, F.G. (1954). *Tables of Integral transforms.* Vol. I, II, McGraw-Hill, New York.

ERDÉLYI, A., MAGNUS, W., OBERHETTINGER, F. AND TRICOMI, F.G. (1955). *Higher transcendental functions,* Vol. III, McGraw-Hill, New York.

EXTON, H. (1972). On two multiple hypergeometric functions related to Lauricella's F_D. *Jñānabha,* Sect. A, 2, 59–73.

EXTON, H. (1972). Certain hypergeometric functions of four variables. *Bull. Soc. Math. Grece,* 13, 104–113.

FETTIS, H.E. (1957). Lommel-type integrals involving three Bessel functions. *J. Math. Phys.* 36, 88–95.

FIELDS, J.L. (1973). Uniform asymptotic expansions of certain classes of Meijer G-functions for a large parameter. *SIAM J. Math. Anal.* 4, 482–507.

FOX, CHARLES (1927). The expression of hypergeometric series in terms of similar series. *Proc. London Math. Soc.* 26 (2) 201–210.

FOX, CHARLES (1928). The asymptotic expansion of generalized hypergeometric functions. *Proc. London Math. Soc.* 27 (2) 389–400.

FOX, CHARLES (1961). The G and H-functions as symmetrical Fourier kernels. *Trans. Amer. Math. Soc.* 98, 395–429.

FOX, CHARLES (1963). Integral transforms based upon fractional integration. *Proc. Cambridge Philos. Soc.* 59, 63–71.

FOX, CHARLES (1965). A formal solution of certain dual integral equations. *Trans. Amer. Math. Soc.* 119, 389–398.

FOX, CHARLES (1965a) A family of distributions with the same ratio property. *Can. Math. Bull.* 8, 631–635.

Fox, Charles (1971). Solving integral equations by L and L^{-1} operators. *Proc. Amer. Math. Soc.* 29, 299–306.

George, A. and Mathai, A.M. (1975). A generalized distribution for the interlivebirth interval. *Sankhya* Ser. B. 37, 332–340.

Gokhroo, D.C. (1970). The Laplace transform of the product of Meijer's G-functions. *Univ. Nac. Tucumán Rev.* Ser. A. 20, 59–62.

Golas, P.C. (1968). Integration with respect to parameters. *Vijnana Parishad Anusandhan Patrika*, 11, 71–76.

Golas, P.C. (1969). On a generalized Stieltjes transform. *Proc. Nat. Acad. Sci.* India Sect. A, 39, 42–48.

Goyal, A.N. (1969). Some infinite series of H-functions. I. *Math. Student* 37, 179–183.

Goyal, A.N. and Goyal, G.K. (1967). On the derivatives of the H-function. *Proc. Nat. Acad. Sci.* India Sect. A, 37, 56–59.

Goyal, A.N. and Goyal, G.K. (1967a). Expansion theorems of H-function. *Vijnana Parishad Anusandhan Patrika* 10, 205–217.

Goyal, A.N. and Chaturvedi, K.K. (1971). Integrals involving Fox H-function. *Univ. Studies* 1, 7–13.

Goyal, A.N. and Sharma. S. (1971). Study of Meijer's G-function of two variables II. *Univ. Studies Math.* 1, 82–89.

Goyal, A.N. and Sharma, S. (1971a). Series of Meijer's G-function of two variables I. *Univ. Studies Math.* 1, 29–35.

Goyal, G.K. (1971). A generalized function of two variables I. *Univ. Studies Math.* 1, 37–46.

Goyal, S.P. (1970). On some finite integrals involving generalized G-function. *Proc. Nat. Acad. Sci.* India Sect. A, 40, 219–228.

Goyal, S.P. (1971). On transformations of infinite series of Fox's H-function. *Indian J. Pure Appl. Math.* 2, No. 4, 684–691.

Goyal, S.P. (1971a). On some finite integrals involving Fox's H-function. *Proc. Indian Acad. Sci.* Sect. A, 74, 25–33.

Goyal, S.P. (1975). The H-function of two variables. *Kyungpook Math.* J. 15, 117–131.

Gulati, H.C. (1971). Fourier series for G-function of two variables. *Gaz. Mat.* (Lisboa) 32, No. 121–124; 21–30.

Gulati, H.C. (1971a). Some contour integrals involving G-function of two variables. *Defence Sci. J.* 21, 39–42.

Gulati, H.C. (1971b). Some formulae for G-function of two variables involving Legendre functions. *Vijnana Parishad Anusandhan Patrika* 14, 77–88.

Gulati, H.C. (1971c). Some recurrence formulae for G-function of two variables I, II. *Defence Sci. J.* 21, 101–106 and 235–240.

Gulati, H.C. (1972). Derivatives of G-function of two variables. *Math. Education* 6, A, 72-A, 76.

Gupta, K.C. (1965). On the H-function. *Ann. Soc. Sci.* Bruxelles Ser. I, 79, 97–106.

Gupta, K.C. (1966). Integrals involving the H-function. *Proc. Nat. Acad. Sci.* India Sect. A, 36, 504–509.

Gupta, K.C. and Jain, U.C. (1966). The H-function II. *Proc. Nat. Acad. Sci.* India Sect. A, 36, 594–609.

Gupta, K.C. and Jain, U.C. (1968a). On the derivative of the H-function. *Proc. Nat. Acad. Sci.* India Sect. A, 38, 189–192.

Gupta, K.C. and Jain, U.C. (1969). The H-function IV. *Vijnana Parishad Anusandhan Patrika*, 12, 25–30.

Gupta, K.C. and Mittal, P.K. (1970). The H-function transform. *J. Austral. Math. Soc.*, 11, 142–148.

GUPTA, K.C. AND MITTAL, P.K. (1971). The H-function transform II, *J. Austral. Math. Soc.*, 12, 444–450.

GUPTA, K.C. AND OLKHA, G.S. (1969). Integrals involving products of generalized hypergeometric functions and Fox's H-function. *Univ. Nac. Tucumàn Rev.* Ser. A, 19, 205–212.

GUPTA, K.C. AND SAXENA, R.K. (1964). Certain properties of generalized Stieltjes transform involving Meijer's G-function. *Proc. Nat. Inst. Sci.* India Sect. A, 30, 707–714.

GUPTA, K.C. AND SAXENA, R.K. (1964). On Laplace transform. *Riv. Mat.* Univ. Parma, Italie, 5, 159–164.

GUPTA, K.C. AND SRIVASTAVA, A. (1970). On certain recurrence relations. *Math. Nachr.* 46, 13–23.

GUPTA, K.C. AND SRIVASTAVA, A. (1971). On certain recurrence relations II. *Math. Nachr.* 49, 187–197.

GUPTA, K.C. AND SRIVASTAVA, A. (1972). On finite expansions for the H-function. *Indian J. Pure Appl. Math.* 3, 322–328.

GUPTA, K.C. AND SRIVASTAVA, A. (1973). Certain results involving Kampé de Feriet's function. *Indian J. Math.* 15, 99–102.

GUPTA, L.C. (1970). Some expansion formulae for Meijer's G-function. *Univ. Nac. Tucumàn Rev.* Ser. A, 20, 109–115.

GUPTA, P.M. AND SHARMA, C.K. (1972). On Fourier series for Meijer's G-function of two variables. *Indian J. Pure Appl. Math.* 3, 1073–1077.

GUPTA, S.C. (1969). Integrals involving products of G-functions. *Proc. Nat. Acad. Sci.* India Sect. A, 39 (2), 193–200.

GUPTA, S.C. (1969a). Reduction of G-function of two variables. *Vijnana Parishad Anusandhan Patrika* 12, 51–59.

GUPTA, S.D. (1973). Some infinite series for the H-function of two variables. *An. Sti. Univ. "Al. I. Cuza"* Iasi Sect. Ia Mat. (N.S.) 19, 185–189.

GUPTA, S.D. (1973a). Fourier series for the H-function of two variables. *An. Sti. Univ. "Al. I. Cuza"* Iasi Sect. Ia Mat. (N.S.) 19, 179–184.

HAHN, W. (1949). Bertrage zur theorie der Heinischen Reihen, die 24 integrale de hypergeometrischen q-differenzengleichung, des q-analogen der Laplace Transformation. *Math. Nachr.* 2, 263–278.

HERZ, C.S. (1955). Bessel functions of matrix argument. *Ann. Math.* 61, 474–523.

HUA, LOO-KENG (1959). *Harmonic analysis of functions of several complex variables in classical domains.* Moscow (In Russian).

JAIN, N.C. (1971). Integrals that contain hypergeometric functions and the H-function. *Repub. Venezuela Bol. Acad. Ci. Fis. Mat. Natur.* 31, No. 90, 95–102.

JAIN, N.C. (1971a). An integral involving the generalized function of two variables. *Rev. Roumaine Math. Pures Appl.* 16, 865–872.

JAIN, P.C. AND SHARMA, B.L. (1968). An expansion for generalized function of two variables. *Univ. Nac. Tucumàn. Rev.* Ser. A, 18, 7–15.

JAIN, P.C. AND SHARMA, B.L. (1968a). Some new expansions of the generalized function of two variables. *Univ. Nac. Tucumàn Rev.* Ser. A, 18, 25–33.

JAIN, R.N. (1965). Some infinite series of G-functions. *Math. Japon.* 10, 101–105.

JAIN, R.N. (1969). General series involving H-functions. *Proc. Cambridge Philos. Soc.* 65, 461–465.

JAIN, U.C. (1967). Certain recurrence relations for the H-function. *Proc. Nat. Inst. Sci.* India Part A, 33, 19–24.

JAIN, U.C. (1968). On an integral involving the H-function. J. *Austral. Math. Soc.* 8, 373–376.

JAISWAL, N.K. (1968). *Priority Queues.* Academic Press, New York.

JAMES, A.T. (1954). Normal multivariate analysis and the orthogonal group. *Ann. Math. Statist.* 25, 40–75.

JAMES, A.T. (1960). The distribution of the latent roots of the covariance matrix. *Ann. Math. Statist.* 31, 151–158.

JAMES, A.T. (1961). The distribution of non-central means with known covariance. *Ann. Math. Statist.* 32, 874–882.

JAMES, A.T. (1961a). Zonal polynomials of the real positive definite symmetric matrices. *Ann. Math.* 74, 456–469.

JAMES, A.T. (1964). Distributions of matrix variates and latent roots derived from normal samples. *Ann. Math. Statist.* 35, 475–501.

JAMES, A.T. (1966). Inference on latent roots by calculations of hypergeometric functions of matrix argument, *Multivariate Analysis*, (P.R. Krishniah, Ed.), pp. 209–235.

JAMES, A.T. (1969). Tests of equality of latent roots of the covariance matrix. *Multivariate analysis* Vol. 2, (P.R. Krishniah, Ed.), pp. 205–218.

JAMES, ALAN T. AND CONSTANTINE, A.G. (1974). Generalized Jacobi polynomials as spherical functions of the Grassman manifold. *Proc. London. Math.* Soc. (3), 29, 174–192.

KALIA, R.N. (1971). An application of a theorem on H-function. *An. Univ. Timi Soara Ser. Sti. Mat.* 9, 165–169.

KALLA, S.L. (1967). Some infinite integrals involving generalized hypergeometric functions ψ_2 and F_c. *Proc. Nat. Acad. Sci.* India Sect. A, 37, 195–200.

KALLA, S.L. (1969). Integral operators involving Fox's H-function. *Acta Mexicana Ci. Tecn.* 3, 117–122.

KALLA, S.L. (1969a). Infinite integrals involving Fox's H-function and confluent hypergeometric functions. *Proc. Nat. Acad. Sci.* India 39A, 3–6.

KALLA, S.L. (1972). An integral involving Meijer's G-function and generalized function of two variables. *Univ. Nac. Tucumàn. Rev. Ser.* A, 22, 57–61.

KALLA, S.L. AND KUSHWAHA, R.S, (1970). Production of heat in an infinite cylinder. *Acta Mexicana Cien. Tecn.* 4, 89–93.

KALLA, S.L. AND MUNOT, P.C. (1970). An expansion formula for the generalized Fox's function of two variables. *Repub. Venezuela Bol. Acad. Ci. Fis. Mat. Natur.* 30, No. 86, 87–93.

KALLA, S.L. AND SAXENA, R.K. (1969). Integral operators involving hypergeometric functions. *Math. Z.* 108, 231–234.

KALLA, S.L. AND SAXENA, R.K. (1971). Relations between Hankel and hypergeometric function operators. *Univ. Nac. Tucumàn Rev.* Ser. A, 21, 231 234.

KAPOOR, V.K. AND GUPTA, S.K. (1970). Fourier series for H-function. *Indian J. Pure Appl. Mat.* 1, No. 4. 433–437.

KAPOOR, V.K. AND MASOOD, S. (1968). On a generalized L-H transform. *Proc. Cambridge Philos. Soc.* 64, 399–406.

KARLSSON, P.W. (1973). Reduction of certain generalized Kampé de Fériet functions. *Math. Scand.* 32, 265–268.

KASHYAP, B.R.K. (1966). The double-ended queue with bulk service and limiting waiting space. *Operations Research.* 14, 822–834.

KAUFMAN, H., MATHAI, A.M. AND SAXENA, R.K. (1969). Distributions of random variables with random parameters. *South Afr. Statist. J.* 3, 1–7.

KAUL, C.L. (1972). Fourier series of a generalized function of two variables. *Proc. Indian Acad. Sci.* Sect. A, 75, 29–38.

KAUL, C.L. (1973). Integrals involving a generalized function of two variables. *Indian J. Pure Appl. Math.* 4, No. 4, 364–373.

KAUL, C.L. (1974). On certain integral relations and their applications. *Proc. Indian Acad. Sci, Sect. A*, 79, 56–66.

KHADIA, S.S. AND GOYAL, A.N. (1970). On the generalized function of 'n' variables. *Vijnana Parishad Anusandhan Patrika* 13, 191–201.

KOBER, H. (1940). On fractional integrals and derivatives. *Quart. J. Math.* Oxford Ser. 11, 193–211.

KUIPERS, L. AND MEULENBELD, B. (1957). On a generalization of Legendre's associated differential equation I and II. *Nederl. Akad. Wetensch. Proc.* Ser. A, 60, 436–450.

KUMAR, RAM (1954). Some recurrence relations of the generalized Hankel transform I. *Ganita*, 5, 191–202.

KUMAR, RAM (1955). Some recurrence relations of the generalized Hankel transform II. *Ganita* 6, No. 1 and 2, 39–53.

KUMAR, RAM (1957). Certain infinite series expansions connected with generalized Hankel transform. *Ganita* 8, No. 1, 1–7.

LAURENZI, BERNARD J. (1973) Derivatives of Whittaker functions $W_{k,1/2}$ and $M_{k,1/2}$ with respect to order k, *Math. Comp.* 27, 129–132.

LAWRYNOWICZ, J. (1969). Remarks on the preceding paper of P. Anandani. *Ann. Polon. Math.* 21, 120–123.

LEBEDEV, N.N. (1965). *Special functions and their applications* (translated from Russian). Prentice-Hall, New Jersey.

LOWNDES, J.S. (1964). Note on the generalized Mehler transform. *Proc. Cambridge Philos. Soc.* 60, 57–59.

LUKE, Y.L. (1962). *Integrals of Bessel functions.* McGraw-Hill, New York.

LUKE, Y.L. (1969). *The special functions and their approximations*, Vol. I, II, Academic Press, New York.

MACROBERT, T.M. (1959). Infinite series for E-functions. *Math. Z.* 71, 143–145.

MACROBERT, T.M. (1961). Fourier series for E-functions. *Math. Z.* 75, 79–82.

MACROBERT, T.M. (1962). *Functions of a complex variable*, 5th ed. Macmillan, London.

MACROBERT, T.M. (1962a). Evaluation of an E-function when three of its upper parameters differ by integral values. *Pacific J. Math.* 12, 999–1002.

MACROBERT, T.M. AND RAGAB, F.M. (1962). E-function series whose sums are constants. *Math. Z.* 78, 231-234.

MCNOLTY, F. AND TOMSKY, J. (1972). Some properties of special functions bivariate distributions, *Sankhya* Ser. B. 34, 251–264.

MAGNUS, W. OBERHETTINGER, F. AND SONI, R.P. (1966). *Formulas and theorems for the special functions of mathematical physics* 52, Springer-Verlag, New York.

MAKAK, RAGY, H. AND SIMARY, M.A. (1971). Summations involving G-functions. *Proc. Math. Phys. Soc.* A.R.E. No. 35, 1–7.

MATHAI, A.M. (1970). *Applications of generalized special functions in statistics.* Monograph. Indian Statistical Institute and McGill University.

MATHAI, A.M. (1970a). The exact distribution of a criterion for testing the hypothesis that several populations are identical. *J. Indian Statistical Assoc.* 8, 1–17.

MATHAI, A.M. (1970b). The exact distribution of Bartlett's criterion for testing equality of covariance matrices. *Publ. L' Isup*, Paris, 19, 1–15.

MATHAI, A.M. (1970c). Statistical theory of distribution and Meijer's G-function. *Metron* 28, 122–146.

MATHAI, A.M. (1971). An expansion of Meijer's G-function in the logarithmic case with applications. *Math. Nachr.* 48, 129–139.

MATHAI, A M. (1971a). On the distribution of the likelihood ratio criterion for testing linear hypotheses on regression coefficients. *Ann. Inst. Statist. Math.* 23, 181–197.

MATHAI, A.M. (1971b). An expansion of Meijer's G-function and the distribution of the product of independent beta variates. *S. Afr. Statist. J.* 5, 71–90.

MATHAI, A.M. (1971c) The exact non-null distributions of a collection of multivariate test statistics. *Publ. L' Isup*, Paris 20, No. 1.

MATHAI, A.M. (1972), Products and ratios of generalized gamma variates. *Skandinavisk Aktuarietidskrift*, 55, 193–198.

MATHAI, A.M. (1972a). The exact distributions of three criteria associated with Wilks' concept of generalized variance. *Sankhya* Ser. A, 34, 161–170.

MATHAI, A.M. (1972b). The exact non-central distributions of the generalized variance. *Ann. Inst. Statist. Math.* 24, 53–65.

MATHAI, A.M. (1972c). The exact distribution of a criterion for testing that the covariance matrix is diagonal. *Trab. Estadisticas* 28, 111–124.

MATHAI, A.M. (1972d). The exact distribution of a criterion for testing the equailty of diagonal elements given that the covariance matrix is diagonal. *Trab. Estadistica* 23, 67–83.

MATHAI, A.M. (1973). A few remarks on the exact distributions of likelihood ratio criteria–1, *Ann. Inst. Statist, Math.* 25, 557–566.

MATHAI, A.M. (1973a). A review of the different methods of obtaining the exact distributions of multivariate test criteria. *Sankhya* Ser. A, 35, 39–60.

MATHAI, A.M. (1973b). A few remarks on the exact distributions of certain multivariate statistics–II. *Multivariate statistical inference*, pp. 169–181, North-Holland Publishing Co., Amsterdam and London.

MATHAI, A.M. (1976). Fox's H-function with matrix argument. (Communicated for publication).

MATHAI, A.M. AND RATHIE, P.N. (1970). An expansion of Meijer's G-function and its application to statistical distributions. *Acad. Roy. Belg. Ci. Sci.* (5) 56, 1073–1084.

MATHAI, A.M. AND RATHIE, P.N. (1970a). The exact distribution of Votaw's criterion. *Ann. Inst. Statist. Math.* 22, 89–116.

MATHAI, A.M, AND RATHIE, P.N. (1970b). The exact distribution for the sphericity test. *J. Statist. Res.* (Dacca) 4, 140–159.

MATHAI, A.M. AND RATHIE, P.N. (1975). *Basic concepts in information theory and statistics: axiomatic foundations and applications*. Wiley Eastern Ltd., New Delhi.

MATHAI, A.M. AND SAXENA, R.K. (1966). On a generalized hypergeometric distribution. *Metrika* 11, 127–132.

MATHAI, A.M. AND SAXENA, R.K. (1969). Distribution of a product and the structural setup of densities. *Ann. Math. Statist.* 40, 1439–1448.

MATHAI, A.M. AND SAXENA, R.K. (1969a). Applications of special functions in the characterization of probability distributions. *S. Afr. Statist. J.* 3, 27–34.

MATHAI, A.M. AND SAXENA, R.K. (1971). Extensions of an Euler's integral through statistical techniques. *Math. Nachr.* 51, 1–10.

MATHAI, A.M. AND SAXENA, R.K. (1971a). Meijer's G-function with matrix argument. *Acta Mexicana Ci. Tecn.* 5, 85–92.

MATHAI, A.M. AND SAXENA, R.K. (1971b). A generalized probability distribution. *Univ. Nac. Tucumàn Rev.* Ser. A, 21, 193–202.

MATHAI, A.M. AND SAXENA, R.K. (1972). Expansions of Meijer's G-function of two variables when the upper parameters differ by integers. *Kyungpook Math, J.* 12, 61–68.

MATHAI, A.M. AND SAXENA, R.K. (1973). On linear combinations of stochastic variables. *Metrika* 20 (3), 160–169.

MATHAI, A.M. AND SAXENA, R.K. (1973a). *Generalized hypergeometric functions with applications in statistics and physical sciences.* Springer-Verlag, Lecture Notes Series No, 348, Heidelberg and New York.

MATHUR, A.B. (1973). Integrals involving H-function. *Math. Student* 41, 162–166.

MATHUR, S.L, (1970). Certain recurrence relations for the H-function. *Math. Education* 4, 132–136.

MATHUR, S.N. (1970). Integrals involving H-functions. *Univ. Nac. Tucumàn, Rev.* Ser. A, 20, 145–148.

McLachlan, N.W. (1961). *Bessel functions for engineers*. 2nd ed. Oxford, University Press, London.

Mehra, A.N. (1971). On certain definite integrals involving the Fox's H-function. *Univ. Nac. Tucumàn Rev*. Ser. A, 21, 43–47.

Meijer, C.S. (1940). Über eine Erweiterung der Laplace-transformation. *Nederl. Akad. Wetensch*. Proc. 43, 599–608, 702–711,=Indag. Math. 2, 229–238, 269–278.

Meijer, C.S. (1941). Eine neue Erweiterung der Laplace-Transform. *Nederl. Akad. Wetensch*. Proc. 44, 727–737. =Indag. Math. 3, 338–348.

Meijer, C.S. (1941a). Multiplikations theoreme für die funktion $G_{p,q}^{m,n}(z)$. *Nederl. Akad. Wetensch*. Proc. 44, 1062–1070.

Meijer, C.S. (1946). On the G-function I–VIII. *Nederl. Akad. Wetensch*. Proc. 49, 227–237; 344–356; 457–469; 632–641; 765–772; 936–943; 1063–1072; 1165–1175 =Indag. Math. 8, 124–134; 213–225; 312–324; 391–400; 468–475; 595–602; 661–670; 713–723.

Meijer, C.S. (1952–1956). Expansion theorems for the G-function. I–XI. *Nederl. Akad. Wetensch*.
Proc. Ser. A, 55=Indag. Math. 14, 369–379; 483–487 (1952);
Proc. Ser. A, 56=Indag. Math. 15, 43–49, 187–193, 349–357 (1953);
Proc. Ser, A, 57=Indag. Math. 16, 77–82, 83–91, 273–279 (1954);
Proc. Ser. A, 58=Indag. Math. 17, 243–251, 309–314 (1955);
Proc. Ser. A, 59=Indag. Math. 17, 70–82, (1956).

Mellin, H.J. (1910). Abrip einer einhaitlichen Theorie der Gamma und der Hypergeometrischen Funktionen. *Math. Ann*. 68, 305–337.

Meulenbeld, B. (1958). Generalized Legendre's associated functions for real values of the argument numerically less than unity. *Nederl. Akad. Wetensch*. Proc. Ser. A, 61, 557–563.

Meulenbeld, B. and Robin, L. (1961). Nouveaux resultats relatifs aux functions de Legendre generalisees. *Nederl. Akad. Wetensch*. Proc. Ser. A, 64, 333–347.

Milne-thomson. L.M. (1933). *The calculus of finite differences*. Macmillan, London.

Mittal, P.K. (1971). Certain properties of Meijer's G-function transform involving the H-function. *Vijnana Parishad Anusandhan Patrika* 14, 29–38.

Mittal, P.K. and Gupta, K.C. (1972). An integral involving generalized function of two variables. *Proc. Indian Acad. Sci*. Sect. A, 75, 117–123.

Mourya, D.P. (1970). Analytic continuations of generalized hypergeometric functions of two variables. *Indian J. Pure Appl. Math*. 1, No. 4, 464–469.

Mourya. D.P. (1970a). The generalized hypergeometric functions of two variables, its analytic continuations and asymptotic expansions. Ph. D. Thesis, University of Indore, Indore, India.

Muirhead, R.J. (1975). Expressions for some hypergeometric functions of matrix argument with applications. *J. Multivariate Analysis* 5, 283–293.

Munot. P C. (1972). Some formulae involving generalized Fox's H-functions of two variables. *Portugal Math*. 31 (4), 203–213.

Munot, P.C. and Kalla, S.L. (1971). On an extension of generalized function of two variables. *Univ. Nac. Tucumàn Rev*. Ser. A, 21, 67–84.

Nair, V.C. (1971). On the Laplace transform–I. *Portugal Math*. 30, 57–69.

Nair, V.C. (1972). Differentiation formulae for the H-function–I,. *Math. Student* 40 A, 74–78.

Nair, V.C. (1972a), The Mellin transform of the product of Fox's H-function and Wright's generalized hypergeometric function. *Univ. Studies Math*. 2, 1–9.

Nair, V.C. (1973). Differentiation formulae for the H-function–II. *J. Indian Math. Soc*. (N.S) 37, 329–334.

Nair, V.C. (1973a). Integrals involving the H-function where the integration is with respect to a parameter. *Math. Student* 41, 195–198.

NAIR, V.C. AND NAMBUDIRIPAD, K.B.M. (1973). Integration of H-functions with respect to their parameters. *Proc. Nat. Acad. Sci. India* Sect. A, 43, 321–324.

NAIR, V.C. AND SAMAR, M.S. (1971). An integrals involving the product of three H-functions. *Math. Nachr.* 49, 101–105.

NAIR, V.C. AND SAMAR, M.S. (1971a). The product of two H-functions expressed as a finite integral of the sum of a series of H-functions, *Math. Education*, 5 A, 45–A, 48.

NARAIN, R. (1965). A pair of unsymmetrical Fourier kernels. *Trans. Amer. Math. Soc.* 115, 356–369.

NATH, RAM (1972). On an integral involving the product of three H-functions. *C.R. Acad. Bulgare Sci.* 25, 1167–1169.

NIELSEN, NIELS (1906). *Handbuch der Theorie der Gamma Funktion.* B.G. Teubner, Leipzig.

NIGAM, H.N. (1969). A note on Fox's H-function. *Ganita* 20, No. 2, 47–52.

NIGAM, H.N. (1970). Integral involving Fox's H-function and integral function of two complex variables I. *Ganita* 21, No. 2, 71–78.

NIGAM, H.N. (1972). Integral involving Fox's H-function and integral function of two complex variables II. *Bull. Calcutta Math. Soc.* 64, 1–5.

OLIVER, M.L. AND KALLA, S.L. (1971). On the derivative of Fox's H-function *Acta Mexicana Ci Tecn.* 5, 3–5.

OLKHA, G.S. (1970). Some finite expansions for the H-function. *Indian J. Pure Appl. Math.* 1, No. 3, 425–429.

PANDA, REKHA (1973). Some integrals associated with the generalized Lauricella functions. *Publ. Inst. Math. (Beograd) Nouvelle* Ser. 16 (30) 115–122.

PARASHAR, B,P. (1967). Fourier series for H-functions. *Proc. Cambridge Philos. Soc.* 63, 1083–1085.

PATHAK, R S. (=PATHAK, RAM SHANKER) (1970). Some results involving G-and H-functions. *Bull. Calcutta Math. Soc.* 62, 97–106.

PATHAK, R.S. (1973). Finite integrals involving products of H-function and hypergeometric function. *Progress Math.* Allahabad 7 (1), 45–72.

PATHAK, R.S. AND PRASAD, V. (1972). The solution of dual integral equations involving H-functions by a multiplying factor method. *Indian J. Pure Appl. Math.* 3, 1099–1107.

PATHAN, M.A. (1968). Certain recurrence relations. *Proc. Cambridge Philos. Soc.* 64, 1045–1048.

PENDSE, ASHA (1970). Integration of H-function with respect to its parameters. *Vijnana Parishad Anusandhan Patrika* 13, 129–138.

PILLAI, K.C.S. AL-AMI, S. AND JOURIS, G.M. (1969). On the distributions of the roots of a covariance matrix and Wilks' criterion for tests of three hypotheses. *Ann. Math. Statist.* 40, 2033–2040.

PRASAD, Y.N. AND SHYAM, DHIR RAM (1973). On some double integrals involving Fox's H-function. *Progress Math.* Allahabad, 7, Nr. 2, 13–20.

RAGAB, F.M. (1963). Expansions of Kampé de Fériet's double hypergeometric functions of higher order. *J. Reine Angew. Math.* Vol. 2, 212, 113–119.

RAGAB, F.M. (1967). Infinite series of Kampé de Fériet's double hypergeometric function of higher order. *Rend. Circ. Mat. Palermo* (2) 16, 225–232.

RAGAB, F. M. AND HAMZA, A.M. (1970). Integrals involving E-functions and Kampé de Fériet's function of higher order. *Ann. Mat. Pura Appl.* (4) 87, 11–24.

RAINVILLE, E.D. (1965). *Special Functions.* Macmillan, New York.

RAKESH, S.L. (1973). Integrals involving products of generalized hypergeometric function and generalized H-function of two variables I. *Univ. Nac. Tucumán Rev.* Ser. A, 23, 281–288.

RAKESH, S.L. (1973a). Recurrence relations. *Defence Sci. J.* 23, 79–84.

RAO, C.R. (1973). *Linear statistical inference and its applications*, 2nd ed. Wiley, New York.

RATHIE, C.B. (1956). A theorem in operational calculus and some integrals involving Legendre, Bessel and E-functions. *Proc. Glasgow Math. Assoc.* 2, 173–179.

RATHIE, C.B. (1960). Integrals involving E-functions. *Proc. Glasgow Math. Assoc.* 4, 186–187.

RATHIE, P.N. (1967). Some finite integrals involving F_4 and H-functions. *Proc. Cambridge Philos. Soc.* 63, 1071–1081.

RATHIE, P.N. (1967a). Some finite and infinite series for F_c, F_4, ψ_2 and G-function. *Math. Nachr.* 35, 125–136.

REED, I.S. (1944). The Mellin type of double integral. *Duke Math. J.* 11, 565–572.

ROUX, J.J. (1975). New families of multivariate distributions. (personal communication).

SAHAI, GOPALJI, (1972). An expansion formula for the generalized function of two variables. *Bull. Math. Soc. Sci. Math. R.S. Roumanie* (N.S.) 16 (64), No. 1, 83–92.

SAKSENA, K.M. (1967). An inversion theory for the Laplace integral. *Nieuw Arch. Wisk.* (3) 15, 218–224.

SAMAR, M.S, (1973). Integrals involving the H-functions, the integration being with respect to a parameter. *J. Indian Math. Soc.* 37, 323–328.

SAMAR, M.S. (1974). Double integrals involving the product of Bessel, F and H-functions. *Vijnana Parishad Anusandhan Patrika* 14, 89–95.

SANSONE, G. AND GERRETSEN (1960). *Lectures on the theory of functions of a complex variable* I. Groningen.

SARAN, S. (1954). Hypergeometric functions of three variables. *Ganita* 5, 77–99.

SARAN, S. (1965). A definite integral involving the G-function. *Nieuw Arch. Wisk* (2) 13, 223–229.

SAXENA, R.K. (1960). Some theorems on generalized Laplace transform -I. *Proc. Nat. Inst. Sci.* India Part A, 26, 400–413.

SAXENA, R.K. (1960a). An integral involving G-function. *Proc. Nat. Inst. Sci.* India Part A, 26, 661–664.

SAXENA, R.K. (1961). Some theorems in operational calculus and infinite integrals involving Bessel function and G-functions. *Proc. Nat. Inst. Sci.* India Part A, 27, 38–61.

SAXENA, R.K. (1961a). A definite integral involving associated Legendre function of the first kind. *Proc. Cambridge Philos. Soc.* 57, 281–283.

SAXENA, R.K. (1962). Definite integrals involving G-functions. *Proc. Cambridge Philos. Soc.* 58, 489–491.

SAXENA, R.K. (1963). Some formulae for the G-function. *Proc. Cambridge Philos. Soc.* 59, 347–350.

SAXENA, R.K. (1963a). Some formulae for the G-function—II. *Collect. Math,* 15, 273–283.

SAXENA, R.K. (1964). Integrals involving G-functions. *Ann. Soc. Sci.* Bruxelles Ser. I, 8, 151–162.

SAXENA, R.K. (1964a). Integrals involving products of Bessel functions. *Proc. Glasgow Math. Assoc.* 6, 130–132.

SAXENA, R.K. (1964b). On some results involving Jacobi polynomials. *J. Indian Math. Soc.* (N.S.) 28, 197–202.

SAXENA, R.K. (1966). Integrals involving products of Bessel functions—II. *Monatsh. Math.* 70, 161–163.

SAXENA, R.K. (1966a). An inversion formula for a kernel involving a Mellin-Barnes type integral. *Proc. Amer. Math. Soc.* 17, 771–779.

SAXENA, R.K. (1966b). An integral involving products of G-functions. *Proc. Nat. Acad. Sci.* India Sect. A, 36, 47–48.

SAXENA, R.K. (1966c). On the reducibility of Appell's function F_4. *Canad. Math. Bull.* 9, 215–222.

SAXENA, R.K. (1967). On the formal solution of certain dual integral equations involving H-functions. *Proc. Cambridge Philos. Soc.* 63, 171–178.

SAXENA, R.K. (1967a). On the formal solution of dual integral equations. *Proc. Amer. Math. Soc.* 18, 1–8.

SAXENA, R.K. (1970). Integrals involving Kampé de Fériet function and Gauss's hypergeometric functions. *Ricerca* (Napoli) 2, 21–27.

SAXENA, R.K. (1971). Integrals of products of H-functions. *Univ. Nac. Tucumàn Rev.* Ser. A, 21, 185–191.

SAXENA, R.K. (1971a). Definite integrals involving Fox's H-function. *Acta Mexicana Ci. Tecn.* 5, No. 1, 6–11.

SAXENA, R.K. (1971b). An integral associated with generalized H-function and Whittaker functions. *Acta Mexicana Ci. Tecn* (3) 5, No. 3, 149–154.

SAXENA, R.K. (1973). Integration of certain product associated with Bessel and confluent hypergeometric functions. *Bull. Math. Soc.* R.S. Roumanie, 16 (64), 93–96.

SAXENA, R.K. (1973a). Abelian theorems for the distributional H-transform. *Acta Mexicana Ci Tecn.* 7, 66–76.

SAXENA, R.K. (1974). On a generalized function of n variables. *Kyungpook Math. J.* 14, 255–259.

SAXENA, R.K. (1977). On the H-function of n variables. *Kyungpook Math. J.* 17, 221–226.

SAXENA, R.K. AND KUMBHAT, R.K. (1973). Fractional integration operators of two variables. *Proc. Indian Acad. Sci.* Bangalore 78, No. 4, 177–186.

SAXENA, R.K. AND KUMBHAT, R.K. (1974). Integral operators involving H-function. *Indian J. Pure Appl. Math.* 5, No. 1, 1–6.

SAXENA, R.K. AND KUMBHAT, R.K. (1974a). Dual integral equations associated with H-function. *Proc. Nat. Acad. Sci.* Allahabad, 44, Sect. A, 106–112.

SAXENA, R.K. AND KUMBHAT, R.K. (1974b). A formal solution of certain triple integral equations involving H-function. *Proc. Nat. Acad. Sci.* Allahabad 44, Part II, Sect. A, 153–160.

SAXENA, R.K. KALLA, S.L. AND BORA, S.L. (1971). Addendum to a paper on integral transforms. *Univ. Nac. Tucumàn Rev.* Ser. 21, 289.

SAXENA, R.K. AND KUSHWAHA, R.S. (1972). Certain dual integral equations associated with a kernel of Fox. *Proc. Nat. Acad. Sci.* India, Sect. A, 42, 39–45.

SAXENA, R.K. AND KUSHWAHA, R.S. (1972a). An integral transform associated with a kernel of Fox. *Math. Student* 40, 201–206.

SAXENA, R.K. AND MATHUR, S.N. (1971). A finite series of the H-functions. *Univ. Nac. Tucumàn Rev.* Ser. A, 21, 49–52.

SAXENA, R.K. AND MODI, G.C. (1974). Some expansions involving H-function of two variables. *Comptes Rendus de l'Academie bulgare des Sci.* 27 (2), 165–168.

SAXENA, R.K. AND MODI, G.C. (1975). Generalized H-function as a symmetrical Fourier kernel. *Univ. Nac. Tucumàn Rev.* Ser. A, 25 (1975) (in Press).

SAXENA, R.K. AND SETHI, P.L. (1973). Relations between generalized Hankel and modified hypergeometric function operators. *Proc. Indian Acad. Sci.* 78, No. 6, 267–273.

SAXENA, R.K. AND SETHI, P.L. (1973a). Certain properties of bivariate distributions associated with generalized hypergeometric functions. *Canadian J. Statist.* 1 (2), 171–180.

SAXENA, R.K. AND SETHI, P.L. (1973b). A formal solution of dual integral equations associated with H-function of two variables. *Univ. Nac. Tucumán Rev.* Ser. A, 23, 121–130.

SAXENA, R.K. AND SETHI, P.L. (1975). Applications of fractional integration operators to triple integral equations. *Indian J. Pure Appl. Math.* 6, No. 5, 512–521.

SAXENA, R.K. AND VERMA, R.U. (1976). Series representations of the H-function of two variables. *Boletin Acad. Cien. Fis. Mat. Natur.* (Caracas), 37 (in press).

SAXENA, R.K. (of Kolhapur) (1968). Definite integrals involving self-reciprocal functions. *Proc. Nat. Inst. Sci.* India Sect. A, 34, 326–336.

SAXENA, V.P. (1970). Inversion formulae to certain integral equations involving H-function. *Portugal Math.* 29 (1), 31–42.

SHAH, MANILAL (1969). Some results on Fourier series for H-functions. *J. Natur. Sci. Math.* 9, No. 1, 121–131.

SHAH, MANILAL (1969a). Some results on the H-functions involving the generalized Laguerre polynomial. *Proc. Cambridge. Philos. Soc.* 65, 713–720.

SHAH, MANILAL (1969b). On some results involving H-functions and associated Legendre functions. *Proc. Nat. Acad. Sci.* India Sect. A, 39, 503–507.

SHAH, MANILAL (1969c). On application of Mellin's and Laplace's inversion formulae to H-functions. *Labdev. J. Sci. Tech.* Part A, 7, 10–17.

SHAH, MANILAL (1969d). On some relation of H-functions and Cebysev polynomials of the first kind. *Vijnana Parishad Anusandhan Patrika*, 12, 61–67.

SHAH, MANILAL (1970). Generalized function of two variables and potential about a spherical surface. *J. Natur. Sci. and Math.* 10, 247–268.

SHAH, MANILAL (1970a). Some results involving generalized function of two variables. *J. Natur. Sci. and Math.* 10, 109–124.

SHAH, MANILAL (1971). Expansion formulas for Meijer's G-function of two variables in series of circular functions. *Jñānabha* Sect. A, No. 1, 35–44.

SHAH, MANILAL (1971a). A result on generalized hypergeometric function and generalized Meijer function of two variables. *An. Sti. Univ. "Al. I. Cuza"* Iasi Sect. I a Mat. (N.S.) 17, 331–338.

SHAH, MANILAL (1971b). Some results on generalized functions and their applications. *Proc. Nat. Acad. Sci.* India, Sect. A, 41, 241–255.

SHAH, MANILAL (1971c). Some results involving generalized Meijer functions associated with Gegenbauer (ultraspherical) polynomials. *Indian J. Pure Appl. Math.* 2, No. 3, 387–400.

SHAH, MANILAL (1971d). On Fourier series for generalized Meijer functions of two variables and their applications. *Indian J. Pure Appl. Math.* 2, No. 3, 464–478.

SHAH, MANILAL (1971e). Some results involving a generalized Meijer function. *Mat. Vesnik* 8 (23), 3–16.

SHAH, MANILAL (1972). On some problems of Fox's H-function of two variables and Gegenbauer polynomials. *Istanbul Tek. Univ. Bul.* 25, No. 2, 111–120.

SHAH, MANILAL (1972a). A note on a generalization of Edelstein's theorem on G-functions. *Glasnik Mat.* Ser. III 7 (27), 201–205.

SHAH, MANILAL (1972b). On generalized Meijer's and generalized associated Legendre functions. *Portugal Math.* 31, 57–66.

SHAH, MANILAL (1972c). Generalized Meijer function and temperature in a non-homogeneous bar. *An. Univ Timisoara* Ser. Sti. Mat. 10, 95–101.

SHAH, MANILAL (1972d). Expansion formulae for H-functions in series of trigonometrical functions with their applications. *Math. Student* 40A, 56–66.

SHAH, MANILAL (1972e). On some problems leading to certain results involving generalized Meijer functions of two variables and associated Legendre functions. *Math. Student* 40A, 124–133.

SHAH, MANILAL (1972f). On some results on H-functions associated with orthogonal polynomials. *Math. Scand.* 30, 331–336.

SHAH, MANILAL (1972g). Some results on generalization of Fox's H-functions. *Bull. Soc. Math. Phys. Macedoine* 23, 13-24.

SHAH, MANILAL (1973). Several properties of generalized Fox's H-functions and their applications. *Portugal Math.* 32, 179–199.

SHAH, MANILAL (1973a). On some applications related to Fox's H-function of two variables. *Publ. Inst. Math.* (Beograd) N. Ser. 16 (30), 123–133.

SHAH, MANILAL (1973b). On a generalized Fox's H-function. *Indian J. Pure Appl. Math.* 4, No. 4, 422–427.

SHAH, MANILAL (1973c). A theorem on generalized Meijer function of two variables. *Istanbul Tek Univ. Bul* 26, No. 1, 30–38.

SHAH, MANILAL (1973d). A new generalized theorem on Fox's H-function. *Gac. Mat.* (Madrid) (1) 26, 158–165.

SHARMA, B.L. (SHARMA BHAGIRATH LAL: ALSO SEE ABIODUN, R.F.A.) (1965). On a generalized function of two variables—I. *Ann. Soc. Sci.* Bruxelles Ser. 1, 79, 26–40.

SHARMA, B.L. (1966). Integrals involving hypergeometric function of two variables. *Proc. Nat. Acad. Sci.* India Sect. A, 36, 713–718.

SHARMA, B.L. (1967). Integrals associated with generalized function of two variables. *Mathematica* (Cluj), 361–374.

SHARMA, B.L. (1967a). Integrals involving generalized function of two variables—II. *Proc. Nat. Acad. Sci.* India Sect. A, 37, 137–148.

SHARMA, B.L. (1968). Some formulae for generalized function of two variables. *Math. Vesnik* 5 (20), 43–52.

SHARMA, B.L. (1968a). An integral involving products of G-function and generalized function of two variables. *Univ. Nac, Tucuman Rev.* Ser. A, 18, 17–23.

SHARMA, B.L. (1971). Sum of a series involving Laguerre polynomials and generalized function of two variables. *An. Sti. Univ. "Al. I. Cuza" Iasi, n. Ser.* Sect. (1) 17, 117–122.

SHARMA, B.L. (1971a). Expansion formulae for generalized function of two variables. *Bull. Math. Soc. Sci. Math.* R.S. Roumainie (N.S) 15 (63), No. 2, 237–245.

SHARMA, B.L. (1972). An integral involving products of G-function and generalized function of two variables. *Rev. Mat. Hisp. Amer.* (4) 32, 188–196.

SHARMA, B.L. AND ABIODUN, R.F.A. (1973). New generating functions for the G-function. *Ann. Polon. Math.* 27, 159–162.

SHARMA, C.K. (1972). Fourier series for Fox's H-function of two variables. *Defence Sci. J.* 22, 227–230.

SHARMA, C.K. (1973). On certain finite and infinite summation formulae of generalized Fox H-functions. *Indian J. Pure Appl, Math.* 4, No. 3, 278–286.

SHARMA, C.K. AND GUPTA, P.M. (1972). On certain integrals involving Fox's H-function. *Indian J. Pure Appl. Math.* 3, 992–995.

SHARMA, K.C. (1964). Integrals involving products of G-function and Gauss hypergeometric function. *Proc. Cambridge Pholos. Soc.* 60, 539–542.

SHARMA, K.C. (1965). On an integral transform. *Math. Z.* 89, 94–97.

SHARMA. O.P. (1965). Some finite and infinite integrals involving H-function and Gauss's hypergeometric functions. *Collect. Math.* 17, 197–209.

SHARMA, O.P. (1966). Certain infinite and finite integrals involving H-function and confluent hypergeometric function. *Proc. Nat. Acad. Sci.* India Sect. A, 36, 1023–1032.

SHARMA, O.P. (1968). On the Hankel transformations of H-functions. *J. Math. Sci.* 3, 17–26.

SHARMA, O.P. (1969). On H-function and heat production in a cylinder. *Proc. Nat. Acad. Sci.* India Sect. A, 39, 355–360.

SHARMA, O.P (1972). Certain infinite integrals involving H-function and MacRobert's E-function. *Labdev. J Sci. Tech.* Part A, 10, 9–13

Simary, M.A. (1973). On hypergeometric functions of matrix argument. *Bull. Math. Soc. Sci. Math.* R.S. Roumanie (N.S) 16 (64), No. 1, 111–118.

Singh, F. (1972). Application of E-operator to evaluate a definite integral. *J. Indian Math. Soc.* (N.S) 35, 217-225.

Singh, F. (1972a). On some results associated with a generalized Meijer function. *Math. Student* 40A, 291–296.

Singh, F. (1972b). Integration of certain products involving H-function and double hypergeometric function II. *Math. Student* 40A, 42–55.

Singh, F. (1972c). Application of E-operator in evaluating certain finite integrals, *Defence Sci. J.* 22, 105–112.

Singh, F. and Varma, R.C. (1972). Application of E-operator to evaluate a definite integral and its application in heat conduction. *J. Indian Math. Soc.* (N.S) 36, 325–332.

Singh, N.P. (1973). A definite integral involving generalized Fox's H-function with applications. *Kyungpook Math. J.* 13, 253–264.

Singh, R.P. (1964). A note on Gegenbauer and Laguerre polynomials. *Math. Japon.* 9, 1–4.

Skibiński, P. (1970). Some expansion theorems for the H-function. *Ann. Polon. Math.* 23, 125–138.

Slater, L.J. (1960). *Confluent hypergeometric functions.* Cambridge University Press.

Slater, L.J. (1961). *Generalized hypergeometric series.* Cambridge University Press.

Sneddon, I.N. (1966) *Mixed boundary value problems in potential theory.* North-Holland Publishing Company, Amsterdam.

Srivastava, Aruna and Gupta, K.C. (1970). On certain recurrence relations. *Math. Nachr.* 46, 13–23.

Srivastava, Aruna and Gupta, K.C. (1971). On certain recurrence relations. II. *Math. Nachr.* 49, 187–197.

Srivastava, G.P. (1971). Some new transformations and reducible casses of Appell's double series and their generalizations. *Math. Student* 39, 319–326.

Srivastava, G.P. and Saran, S. (1967). Integrals involving Kampe de Fériet function. *Math. Z.* 98, 119–125.

Srivastava, H.M. (1964). Hypergeometric function of three variables. *Ganita* 15, 97–108.

Srivastava, H.M. (1972). A contour integral involving Fox's H-function. *Indian J. Math.* 14, 1–6.

Srivastava, H.M. (1972a). A class of integral equations involving H-function as kernel. *Nederl. Acad. Wetensch. Proc.* Ser. A 75=Indag. *Math.* 34, 212–220.

Srivastava, H.M. (1973). On the reducibility of Appell's function F_4. *Canadian Math. Bull.* 16, 295–298.

Srivastava. H.M. and Buschman, R.G. (1973). Composition of fractional integral operators involving Fox's H-function. *Acta Mexicana de Ciencia y Technologia* 7, No. 1–2–3, 21–28.

Srivastava, H.M. and Buschman, R.G. (1974). Some convolution integral equations. *Nederl. Akad. Wetensch. Proc.* Ser. A 77 (3)=Indag. Math. 36 (3), 211–216.

Srivastava, H.M. and Buschman, R.G. (1975). Some polynomial defined by generating relations. *Trans. Amer. Math. Soc.* 205, 360–370.

Srivastava, H.M. and Buschman, R.G. (1976). Mellin convolutions and H-function transformations. *Rocky Mountain J. Math.* 6, No. 2, 341–343.

Srivastava, H.M. and Daoust, Martha, C. (1969). On Eulerian integrals associated with Kampé de Fériet's function. *Publ. Inst. Math.* Nouvelle Serie 9 (23), 199–202.

Srivastava, H.M. and Daoust, Martha, C. (1969a). Certain generalized Neumann expansions associated with the Kampé de Fériet function. *Nederl. Akad. Wetensch.*

Proc. Ser. A 72, No. 5=Indag. Math. 31 No. 5, 449–457.

SRIVASTAVA, H.M. AND DAOUST, MARTHA, C. (1973). A note on the convergence of Kampé de Fériet's double hypergeometric series. *Math. Nachr.* 53, 151–159.

SRIVASTAVA, H.M., GUPTA, K.C. AND HANDA, S. (1975). A certain double integral transformation. *Nederl. Akad. Wetensch. Proc.* Ser. A 78=Indag. Math. 37, 402–406.

SRIVASTAVA, H.M. AND JOSHI, C.M. (1968). Certain integrals involving a generalized Meijer function. *Glasnik Mat.* Ser. III 3(23), 183–191.

SRIVASTAVA, H.M. AND JOSHI, C.M. (1969). Integration of certain products associated with a generalized Meijer function. *Proc. Cambridge Philos. Soc.* 65, 471–477.

SRIVASTAVA, H.M. AND PANDA, R. (1973). Some operational techniques in the theory of special functions. *Nederl. Akad. Wetensch. Proc.* Ser. A 76=Indag. Math. 35, 308–319.

SRIVASTAVA, H.M. AND PANDA, R. (1976). Some bilateral generating functions for a class of hypergeomctric polynomials. *J. Reine. Agnew, Math.* 283/284, 265–274.

SRIVASTAVA, H.M. AND SINGHAL, J.P. (1968). Double Meijer transformations of certain hypergeometric functions. *Proc. Cambridge Philos. Soc.* 64, 425–430.

SRIVASTAVA, H.M. AND SINGHAL, J.P. (1969). Certain integrals involving Meijer's G-function of two variables. *Proc. Nat. Inst. Sci.* India Part A, 35, 64–69.

SRIVASTAVA, H.M. AND VERMA, R.U. (1970). On summation of Meijer's G-function of two variables. *Indian J. Math.* 12, 137–140.

SRIVASTAVA, K.N. (1964). Some polynomials related to Laguerre polynomials. *J. Indian Math. Soc.* (N.S.) 28, 43–50.

SRIVASTAVA, M.M. (1969). Infinite series of H-functions. *Istanbul Univ. Fen Fak. Mecm.* Ser. A, 34, 79–81.

SRIVASTAVA, SUNIL KUMAR (1972). Fourier series for H-function of two variables. *Math. Bulkanica* 2, 219–225.

SRIVASTAVA, SUNIL KUMAR (1972a). On the H-function of two variables. *Bull. Math. Soc. Sci. Math.* R.S.R. n. Ser. 16 (64), 119–123.

SRIVASTAVA, T.N. AND SINGH, Y.P. (1968). On Maitland's generalized Bessel function. *Canad. Math. Bull.* 2, 739–741.

SUBRHMANIAM, KOCHERLAKOTA (1973). On some functions of matrix argument. *Utilitas Math.* 3, 83–106.

SUBRAHMANIAM, KOCHERLAKOTA (1974). Recent trends in multivariate normal distribution theory: On the zonal polynomials and other functions of matrix argument. *Technical Report* No. 69, University of Manitoba (Department of Statistics).

SUD, K. AND WRIGHT, L.E. (1976) A new analytic continuation of Appell's hypergeometric series F_2. *J. Mathematical Physics,* 17, 9, 1719–1721.

SUNDARARAJAN, P.K. (1966). On the derivative of a G-function whose argument is a power of the variable. *Compositio Math.* 7, 286–290.

SWAROOP, RAJENDRA (1965). A general expansion involving Meijer's G-function. *Ann. Soc. Sci.* Bruxelles Ser. 1, 79, 47–57.

SZEGÖ, G. (1939). Orthogonal polynomials. *American Math. Soc.* Colloquium Publication No. 23.

TAXAK, R.L. (1970). Some results involving Fox's H-function and associated Legendre functions. *Vijnana Parishad Anusandhan Patrika* 13, 161–168.

TAXAK, R.L. (1971). Integration of some H-functions with respect to their parameters. *Defence Sci. J.* 21, 111–118.

TAXAK, R.L. (1971a). A contour integral involving Fox's H-function and Whittaker function. *An. Vac. Ci.* Univ. Porta 54, 353–362.

TAXAK, R.L. (1971b). Fourier series for Fox's H-function. *Defence Sci. J.* 21, 43–48.

TAXAK, R.L. (1972). Some integrals involving Bessel's functions and Fox's H-function. *Defence Sci. J.* 22, 15–20.

TAXAK, R.L. (1973). Some series for the Fox's H-function. *Defence Sci. J.* 23, 33–36.

TOSCANO, L. (1944). Transformata di Laplace di Prodotti di fanzioni di Bessel polinomi di Laguerre F_A di Louricella. *Part. Acad. Sci.* (commen) (5), 471–500.

TOSCANO, L. (1972). Sui polinomi ipergeometriche a piu variabili del tipo F_D di Lauricella. *Le Matematische* 27, 219–250.

TRANTER, C.J. (1956). *Integral transforms in mathematical physics.* 2nd ed. Methuen, London.

TRANTER, C.J. (1969). *Bessel function with some physical applications.* Hart Publishing Co., New York.

TREMBLAY, R. AND LAVERTU, M.L. (1972). P. Humber's confluent hypergeometric function $\phi_1(\alpha, \beta; \gamma; x, y)$. *Jñānabha,* Sect. A, Vol. 2, 11–18.

VARMA, V.K. (1963). On another representation of an H-function. *Proc. Nat. Acad. Sci.* India, Sect. A (2), 33, 275–278.

VARMA, V.K. (1965). On a multiple integral representation of a kernel of Fox. *Proc. Cambridge Philos. Soc.* 61, 469–474.

VARMA, V.K. (1966). On a new kernel and its relation with H-function of Fox. *Proc. Nat. Acad. Sci.* India Sect. A, (2), 36, 389–394.

VASISHTA, SHIVKANT (1974). Some integrals involving the H-function of two variables. *Math. Education* 8, A, 65-A, 71.

VERMA, ARUN (1966). A note on an expansion of hypergeometric functions of two variables. *Math. Comp.* 20, 413–417.

VERMA, ARUN (1966a). Expansions involving hypergeometric functions of two variables. *Math. Comp.* 20, 590–596.

VERMA, C.B.L. (1966). On H-function of Fox. *Proc. Nat. Acad. Sci.* India Sect. A (3) 36, 637–642.

VERMA, R.U. (1966). Certain integrals involving G-function of two variables. *Ganita* 17, 43–50.

VERMA, R.U. (1966a). On some integrals involving Meijer's G-function of two variables. *Proc. Nat. Inst. Sci.* India, Sect. A, 32, 509-515.

VERMA, R.U. (1967). On some infinite series of the G-function of two variables. *Mat. Vesnik* 4(19), 265–271.

VERMA, R.U. (1969/70). Reduction formula for Meijer's G-function of two variables. *Univ. Lisboa Revista Fac. Ci.* A (2), 13, 131–133.

VERMA, R.U. (1970). Addition theorem on G-function of two variables. *Math. Vesnik* 7 (22), 165–168.

VERMA, R.U. (1970a). Expansion formula for the G-function of two variables. *An. Sti. Univ. "Al. I. Cuza"*, Iasi n. Ser. Sect. Ia, 16, 289–291.

VERMA, R.U. (1971). On the H-function of two variables II. *An. Sti. Univ. "Al. I. Cuza"*, Iasi Sect. Ia, Mat. (N.S.) 17, 103–109.

VERMA, R.U. (1971a). Integrals involving G-function of two variables II. *C.R. Acad. Bulgare Sci.* 24, 427–430.

VERMA, R.U. (1971b). On the H-function of two variables V. *An. Univ. Timisoara Ser. Sti. Mat.* 9, 205–209.

VERMA, R.U. (1972). H-function of two variables VI. *Defence Sci. J.* 22, 241–244.

VERMA, R.U. (1972a). A generalization of integrals involving Meijer's G-function of two variables. *Math. Student* 40A, 40–46.

VERMA, R.U. (1974). Solution of an integral equation by L and L^{-1} operators. *An. Sti. Univ. "Al. I. Cuza"*, Iasi 20, 381–387.

VYAS, R.C. AND SAXENA, R.K. (1973). Integrals involving G-function of two variables. *Univ. Nac. Tucumán. Rev. Sci.* A, 23, 17–23.

VYAS, R.C. AND SAXENA, R.K. (1974). On Kummer's transforms of two variables involving Meijer's G-function. *Rev. Mat. Hisp. Amer.* (4) 34, 335–338.

WRIGHT, E.M. (1935). The asymptotic expansion of the generalized Bessel function. *Proc. London Math. Soc.* (2) 38, 257–270.

WRIGHT, E.M. (1935a). The asymptotic expansion of the generalized hypergeometric function. *J. London Math. Soc.* 10, 286–293.

WRIGHT, E.M. (1940). The asymptotic expansion of the generalized hypergeometric function. *Proc. London Math. Soc.* (2) 46, 389–408.

WRIGHT, E.M. (1940a). The generalized Bessel function of order greater than one. *Quart. Jour. Math.* Oxford Ser. 11, 36–48.

Symbols Index

Author Index

Subject Index